Engineering Economics for the 21st Century

Engineering Economics for the 21st Century

Paul Marnell
Manhattan College

WILEY

Published by John Wiley & Sons, Inc., Hoboken, New Jersey.
Published simultaneously in Canada.

Library of Congress Cataloging-in-Publication Data is available.

ISBN: 978-1-118-92903-2

Printed in the United States of America.

10 9 8 7 6 5 4 3 2 1

Contents

9 Aftertax IRR%

Chapter 10 Sensitivity Analysis

Chapter 11 Other Profitability Measures

Preface

Rational for the Book

Since there is a plethora of texts on engineering economics, why did I choose to write one? An historical perspective provides the answer. Eugene L. Grant wrote the first edition of his pioneering text on engineering economics in 1931, a time when engineers and students had modest mathematics training and computational tools that were limited to slide rules and tables of mathematical functions and engineering economics factors. In the ensuing years numerous texts on the subject have appeared, but the expositions in current engineering economics texts only differ from Grant's last edition in 1990 by the inclusion of topics from *managerial* economics. *The universal reliance on* ***tabulations*** *of 8 of the 12 factors of engineering economics has not changed. In addition, the practice of engineering economics is still constrained by the following* ***seven limitations of*** *past and current* ***engineering economics texts:***

1. Limited use of the important but untabulated geometric gradient series in calculations.
2. The tedious use of the *limited* tabulated factor values in calculations.
3. Using factor equations when tabulated factor values are not available.
4. The tedious practice of manually preparing certain least-cost studies.
5. Defining the **internal rate of return** *IRR* in Chapter 5 or 6 of a past or present text as "the interest rate earned on the unrecovered balance of an investment so that the balance is zero at the end of the investment's life." But *the meaning of "balance" is rarely referenced in the Index of a text in which this ubiquitous IRR definition occurs.* Then it is stated *without proof* that the receipt and disbursement series are equivalent, though the concept of equivalence is often casually defined.
6. *Not discussing the existence of* ***quasi-simple investments*** *and, therefore, not providing the methods for finding their IRRs.*
7. *Not discussing the existence of* ***four classes of practical non-simple investments*** *and, therefore, not providing methods for finding their IRRs.*

The ***goals*** *of this* ***text*** *are* ***(1) the elimination of the seven limitations of engineering economics texts*** *and (2) enhancement of the explanations of the various profitability measures by applying them to the* ***same*** *investment.* These goals can be achieved in a 15-week semester with *three* lectures per week.

To remove Limitations 1, 2, and 3, I designed the ***12-FACTOR Calculator*** spreadsheet in Excel.[1] It calculates the values of the *12-Factors* for *arbitrary* values of $i\%$, N, and $g\%$ when

$$g\% \neq i\%.$$

[1] Excel is a trademark of Microsoft Corporation.

	A	B	C	D	E	F	G	H
1								
2		***12-FACTOR Calculator***						
3				*i%*	*g%*			
4			***Input:***	1.5%	4%			
5			***Number of Interest Periods N***					
6		***Input:***	2	4	6	8	10	12
7			***Solution: Present Worth Factor Values***					
8		$(P/F, i, N)$	0.97066	0.94218	0.91454	0.88771	0.86167	0.83639
9		$(P/A, i, N)$	1.95588	3.85438	5.69719	7.48593	9.22218	10.90751
10		$(P/G, i, N)$	0.97066	5.70985	13.99560	25.61574	40.36748	58.05708
11		$(P/A_1, g, i, N)$	1.99471	4.08889	6.28751	8.59576	11.01912	13.56333
12			***Solution: Annual Worth Factor Values***					
13		$(A/P, i, N)$	0.51128	0.25944	0.17553	0.13358	0.10843	0.09168
14		$(A/F, i, N)$	0.496278	0.24444	0.16053	0.11858	0.09343	0.07668
15		$(A/G, i, N)$	0.49628	1.48139	2.45658	3.42185	4.37721	5.32267
16		$(A/A_1, g, i, N)$	1.01985	1.06084	1.10362	1.14826	1.19485	1.24349
17			***Solution: Future Worth Factor Values***					
18		$(F/P, i, N)$	1.03023	1.06136	1.09344	1.12649	1.16054	1.19562
19		$(F/A, i, N)$	2.01500	4.09090	6.22955	8.43284	10.70272	13.04121
20		$(F/G, i, N)$	1.00000	6.06022	15.30340	28.85594	46.84811	69.41410
21		$(F/A_1, g, i, N)$	2.05500	4.33980	6.87503	9.68306	12.78814	16.21656

1. The ***12-FACTOR Calculator*** provides the values of the *12-Factors* for *arbitrary* values of *i%*, *N*, and *g%* when $g\% \neq i\%$. Since some problems require factors with several values of *N*, the ***12-FACTOR Calculator*** provides six *input* cells for them.
2. The ***12-FACTOR Calculator*** significantly reduces the computation time of problems and elevates the accuracy of their solutions by eliminating factor table lookups and minimizing reading errors.
3. The ***12-FACTOR Calculator*** eliminates the need to lookup factor equations to compute factor values that are not tabulated.
4. The ***12-FACTOR Calculator*** allows the practical use of the geometric gradient series to represent a cost component of a cost alternative or represent wholly or in part the receipt and/or disbursement series of a loan or investment.

Limitation 4 is eliminated by *automating* the annual or present worth least-cost studies required for the purchase of business and industrial equipment. To do so, I created the following Excel spreadsheets: ***ANNUAL WORTH COMPARISON Calculator with COMPONENT REPLACEMENT*** and the ***PRESENT WORTH COMPARISON Calculator with COMPONENT REPLACEMENT***. They are displayed on the following page. The variables C_0, l, and $f\%$ represent the initial cost, life, and annual percentage change in the cost of a component that is replaced during service time *ST* of the study.

	A	B	C	D	E	F	G	H
1								
2		***ANNUAL WORTH COMPARISON Calculator with COMPONENT REPLACEMENT***						
3		***Input:***				***Solution: Annual Worths and Sums***		
4			***X***	***Y***	***Z***	***X***	***Y***	***Z***
5		***P***	$165,000	$185,000	$200,000	$39,356	$44,127	$47,705
6		***S ⊕ MV***	($11,800)	($23,000)	($25,400)	($455)	($886)	($978)
7		C_0	$34,000	$28,000	$22,000			
8		l	4	5	5	$6,791	$3,265	$2,445
9		$f\%$	3%	4%	3%			
10		$\#_{max} = 2$	1	0	0			
11		***A***	$4,000	$1,800	$1,200	$4,000	$1,800	$1,200
12		***G***	$250	$175	$150	$768	$538	$461
13		A_1	$0	$300	$0	$0	$281	$0
14		$g\%$	0%	4%	0%			
15		$A_{1,net}$	0	1	0			
16		$r^*\%$	20.00%		***ΣAW***	**$50,460**	**$49,125**	**$50,833**
17		***ST***	10					

	A	B	C	D	E	F	G	H
1								
2		***PRESENT WORTH COMPARISON Calculator with Component Replacement***						
3		***Input:***				***Solution: Present Worths and Sums***		
4			***X***	***Y***	***Z***	***X***	***Y***	***Z***
5		***P***	$165,000	$185,000	$200,000	$165,000	$185,000	$200,000
6		***S ⊕ MV***	($11,800)	($23,000)	($25,400)	($1,906)	($3,715)	($4,102)
7		C_0	$34,000	$28,000	$22,000			
8		l	4	5	5	$28,471	$13,690	$10,249
9		f	3%	4%	3%			
10		$\#_{max} = 2$	1	0	0			
11		***A***	$4,000	$1,800	$1,200	$16,770	$7,546	$5,031
12		***G***	$250	$175	$150	$3,222	$2,255	$1,933
13		A_1	$0	$300	$0	$0	$1,177	$0
14		$g\%$	0%	4%	0%			
15		$A_{1,net}$	0	1	0			
16		$r^*\%$	20.00%		***ΣPW***	**$211,557**	**$205,953**	**$213,111**
17		***ST***	10					

The elimination of **Limitation 5**, the vague definition of the IRR in Chapter 5 or 6 of contemporary texts, begins on page 22 of **Chapter 2**. The **Balance Equation** is presented and explained:[2]

$$\boxed{R_k - P_k = B_{k-1}(1+i) - B_k \qquad k = 1, 2, \ldots, N.} \tag{1}$$

Next the **General Repayment Equation** (**GRE**) for any loan or investment is derived from Equation (1):

$$\boxed{\sum_{k=0}^{N} \frac{R_k}{(1+i)^k} = \sum_{k=0}^{N} \frac{P_k}{(1+i)^k}.} \tag{2}$$

Hence, ***the present worth of a loan's principles (or an investment's disbursements) equals the present worth of the loan's payments (or an investment's receipts).*** That is, ***the two series are equivalent series.***

When Equation (2) is applied to a proposed investment, everything is known except the rate of return, which is denoted by i^*. Substituting the definitions for the kth cash flow $CF_k \equiv R_k - D_k$ and $x \equiv 1 + i^*$ into Equation (2) yields the **General Polynomial Equation** (GPE) for the unknown value of x and, hence, i^*:

$$\boxed{CF_0 x^N + CF_1 x^{N-1} + \cdots + CF_{N-1} x + CF_N = 0.} \tag{3}$$

When we define a **simple investment** as an investment with one change in the signs of its cash flows, Descartes' Rule of Signs tells us and our students that there is only one positive root of the GPE. Hence, there is only one value of i^* in its range of permissible values:

$$-1 < i^* < \infty.$$

(A separate derivation in Chapter 6 shows that *all* of the *intermediate balances* of a simple investment are *positive*.)

Excel's IRR function is explained and used in *Chapter 2* to find the IRR of a simple investment, and it is used in *Chapter 3* to find the IRR of a *spec* house (a house built without a buyer) with a series of 14 *monthly* cash flows. The resulting monthly IRR is converted to an Annual Percentage Yield *APY*. The IRR function is also used in *Chapter 3* to find the IRR of a *proposed* investment and determine a counter offer that will provide a desired IRR.

The bottom line is that ***finding the IRR of an investment is not based on the vague definition of the IRR in a typical contemporary text. It is precisely defined in Chapter 2 and used wherever a need for the IRR arises in the book.***

Limitation 6, *not recognizing the existence of quasi-simple investments, is covered in Chapter 6.* A **multi-simple investment** is *defined* as a sequence of two or more simple investment patterns. The cash flow diagram of the *tri-simple* investment shown on the following page has three simple investment patterns. Step-by-Step 6.1 is used to determine *if* a multi-simple investment has a *unique* value of i^* and *positive* intermediate balances. Two examples illustrate the simple calculations for finding the *IRR*s of bi-simple and tri-simple investments.

---

[2] A simpler form is used earlier in the chapter to derive the ***fundamental* repayment *equation***:

$$F = P(1+i)^n.$$

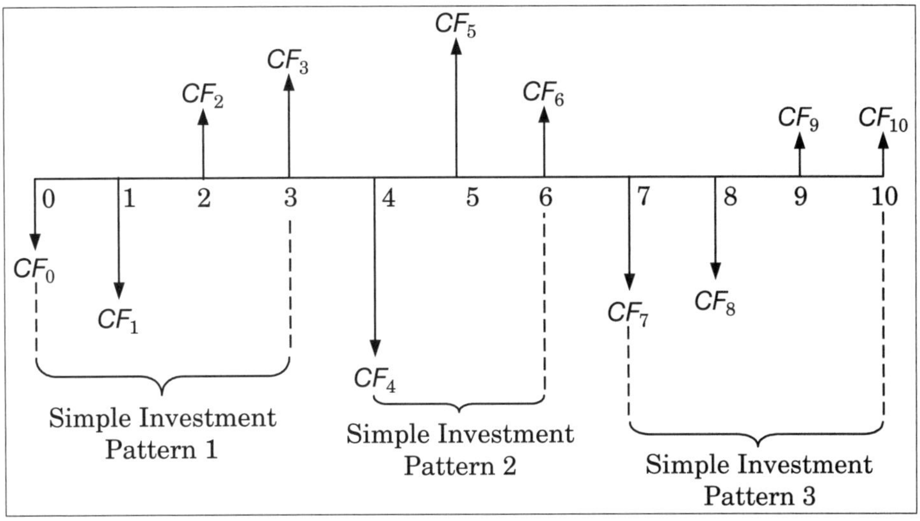

CFD of a Tri-Simple Investment Note: There are two negative cash flows in the first and third simple investment patterns.

Limitation 7, not providing an *IRR* solution strategy for the *four basic* non-simple investments, is covered in the ***comprehensive 27-page Chapter 7***. It begins with a review of the five fundamental properties of simple and quasi-simple investments. They are used to identify the properties that are *not* possessed by the **four basic non-simple investments** listed in Table 7.1.

Table 7.1 Properties of the Four Basic Non-Simple Investments

FUNDAMENTAL PROPERTIES of a SIMPLE INVESTMENT	PROPERTIES of the FOUR BASIC NON-SIMPLE INVESTMENTS			
	DOWN PAYMENT	TERMINATION	DOWN PAYMENT & TERMINATION	MULTI-SIMPLE
NEGATIVE FIRST CASH FLOW	✗	✔	✗	✔
POSITIVE LAST CASH FLOW	✔	✗	✗	✔
ODD NUMBER of CASH FLOW SIGN CHANGES (1)	✗ (2)	✗ (2)	✔ (3)	✔ $(2N-1)^1$
UNIQUE RATE of RETURN	✗	✗	Possible	Possible
POSITIVE INITIAL and INTERMEDIATE BALANCES	✗	✗	✗	Possible

[1] N is the number of simple investment patterns in a multi-simple investment.

The meaning of a negative balance is explained, and the need for an ***external investment*** for the negative balances is presented. *This is not new.* ***What is new***

is the presentation of three easy to use Step-by-Steps for calculating the IRRs of the four classes of non-simple investments *listed in Table 7.1. Chapter 7 has six numerical examples that illustrate these calculations.*

Some practitioners say that non-simple investments rarely occur in practice. Table 7.2 shows that this is an *overstatement*.

Table 7.2 Occurrences of Non-Simple Investments

NON-SIMPLE INVESTMENT	OCCURRENCE
DOWN PAYMENT	*Any* investment where a down payment, an initial positive cash flow, is required before a contractor will initiate a project that consists of one or more subsequent negative cash flows followed by one or more positive cash flows.
TERMINATION	A termination investment often occurs when an initially simple investment encounters problems, such that one or more *unexpected* negative cash flows occur at the *end* of the project. For example, a chemical plant that must clean up its site because it is polluting a town water supply will often have one or more *project-ending* negative cash flows.
DOWN PAYMENT & TERMINATION	This type of investment is usually associated with the lease of an oil field. *Initially* there will be one or more positive cash flows followed by a negative cash flow to repressurize the oil field. The negative cash flow is followed by more positive cash flows and one or more terminating negative cash flows to restore the land to its natural state.
MULTI-SIMPLE	This type of investment often occurs when a simple investment encounters a large intermediate negative cash flow due to a strike, fire and explosion, or other unforeseen negative occurrence.

In addition to eliminating the seven limitations of contemporary engineering economics texts, this book enhances the explanations of (A) **Pretax IRR**, (B) **Aftertax IRR**, (C) **Sensitivity Analysis**, and (D) **Other Profitability Measures** by illustrating them with a **common** investment: a *Chemical Plant Investment* **CPI**.

TABLE 6.1 Values of the Economic Variables of a *Proposed* CPI

ECONOMIC VARIABLE	*ESTIMATED* VALUE
Fixed Capital: FC	$\$69.6 \times 10^6$
Land Capital: $LC \approx 0.01FC$	$\$0.696 \times 10^6$
Working Capital: $WC \approx 0.15FC$	$\$10.44 \times 10^6$
Salvage Value: $S \approx 0.05FC$	$\$3.48 \times 10^6$
Annual Receipt: R	$\$208.8 \times 10^6$
Annual Production Cost: PC	$\$147 \times 10^6$
Annual SCORD Cost: $SC \approx 0.15R$	$\$31.32 \times 10^6$
Capital Time: CT	2 a
Operating Time: OT	10 a

Spreadsheet 6.1 below, which is on page 136, calculates a *pretax* IRR% of 32.59% for the **CPI**.

	A	B	C	D	E
1					
2		***Pretax IRR% of a CHEMICAL PLANT INVESTMENT: CT = 2a***			
3			***FC***	$69,600,000	= 69.6*10^ 6
4			***LC***	$696,000	= 0.01*D3 These are typical ***cost estimating equations.***
5			***WC***	$10,440,000	= 0.15*D3
6			***S***	$3,480,000	= 0.05*D3
7			***R***	$208,800,000	= 208.8*10^ 6
8			***PC***	$147,000,000	= 147* 10^ 6
9			***SC***	$31,320,000	= 0.15*D7 This is a ***cost stimating equation.***
10			***k***		**Cash Flow, $10^6**
11			0	$-LC$	(0.696)
12			1	$-\frac{1}{3}FC$	(23.2)
13			2	$-\left(\frac{2}{3}FC + WC\right)$	(56.84)
14			3		30.48
15			4		30.48
16			5		30.48
17			6		30.48
18			7	$OCF = R - (PC + SC)$	30.48
19			8		30.48
20			9		30.48
21			10		30.48
22			11		30.48
23			12	$OCF + LC + WC + S$	45.096
24				**= IRR(E11:E23)**	**32.59%**

Spreadsheet 9.1 on page 183 shows that the payment of federal and state income taxes reduces the IRR% to 23.81%, and Spreadsheet 10.2 on page 200 has a **Data Table** that shows how the IRR% of the **CPI** varies with the annual receipt *R*. Figure 10.1 on page 201 is the **sensitivity curve** for this relationship.

A Few Other Unique Features

Chapter 3 (1) Intermediate loan and investment balances are calculated *recursively*. **(2)** Spreadsheet 3.1 calculates the yearly balances and annual interests of a 30-year mortgage, and Figure 3.1 shows the variation of the normalized annual interests over the life of the mortgage. **(3)** The Composite Series Transformation Procedure for finding the present worths of a composite series is explained and illustrated. **Chapter 6 (1)** The End-of-Year (E-o-Y) Convention is carefully defined, and its application to an investment with a fixed capital **S-Curve** and a conventional operating time is illustrated. **(2)** The engineering, purchasing, construction, and

start-up phases of a **CPI** are briefly explained. **(3)** The *pretax* IRR% of the **CPI** is calculated.
Chapter 8 presents a detailed comparison of six ***mutually exclusive*** plant sites.
Chapter 9 presents spreadsheet calculations of (1) the *aftertax* IRR% of the **CPI**, (2) the effect of a *delayed* startup on its IRR%, and (3) a case where straight-line depreciation is *advantageous.*
Chapter 10 explains how to create a **Data Table** in Excel to determine how changes in the value of a variable, such as the annual receipt R of the **CPI**, affects the IRR%. Spreadsheet 10.1 illustrates the calculation for the **CPI**, and Figure 10.1 is a graph, the **sensitivity curve**, of the data.
Chapter 11 applies the other profitability methods, **net present value**, **accounting rate of return**, and **benefit-cost ratio** to the **CPI**.

The *Problem Solutions* files, and the ***12-CUSTOM Factor Calculator, ANNUAL WORTH COMPARISON Calculators, PRESENT WORTH COMPARISON Calculators,*** and the ***IRR spreadsheets*** for the **CPI** are available at ***booksupport.wiley.com*** by entering ***ISBN 9781118929032.***

The ***pedagogic goals*** of the book are the following:

1. Increase Student Computational Skills
2. Eliminate Tedious Trial and Error Calculations
3. Enhance Student Comprehension

The various ***Calculators*** and ***IRR spreadsheets*** provided with the book achieve Goals 1 and 2. Goal 3 is achieved as follows:

All chapters illustrate the use of some or all of following items:

1. Different typefaces are used to identify **definitions** and *important words* and *results.*
2. Various types of italicized notes, such as *Computation Notes* and *Usage Notes*, explicitly highlight important material.
3. *Important equations*, such as the **recursive balance equation**, are boxed:

$$\boxed{B_k = B_{k-1}(1+i) - CF_k \qquad k = 1, 2, \ldots, N.}$$

4. Carefully **annotated spreadsheets** identify input variables, the contents of value cells, and the commands used in formula cells.

*I use typography, shading, and various Notes so that the student can concentrate on understanding the important material instead of trying to figure out **what is** the important material!*

Dedications

This book is dedicated to the women in my life: Patricia, my incredible wife, who provided timeless support on this literary journey; Vanessa, my do-it-all daughter, for her encouragement and always keeping me on-track; Samantha, my darling two-year old granddaughter, who provides so many fun times and so much laughter; Leona, my mother-in-law and fan, I wish she could be here to throw the publication party she always spoke of and to receive her signed copy of the book.

1

Overview of Engineering Economics

"*...engineering...is the art of doing that well with one dollar, which any bungler can do with two after a fashion.*"
Arthur Mellon Wellington, 1889

"*It is apparent that the so-called principles of design are subordinate to the principles that underlie economic judgment.*"
J.C.L. Fish

1.1 WHY STUDY ENGINEERING ECONOMICS?

• Companies are **money factories** that convert raw materials and/or brain power into dollars.
• The success of any company that you work for depends on it being able to provide a product or service that can be sold at a profit that yields an acceptable **rate of return**. This is its **mission**.
• Regardless of your job description, completion of your tasks in a *timely* and *cost-effective* manner contributes to the achievement of your company's mission.
• Choosing between engineering alternatives, such as three heat exchangers or two office building designs, is based on technical and **economic** evaluations of the alternatives.
• The decision to fund a technical project, such as the construction of a chemical plant, is based on a technical evaluation of the project and its *anticipated* **rate of return**.
• You ***can also use the principles of engineering economics to***
  Decide whether it's cheaper to rent or buy a house.
  Determine the monthly cost of leasing a car.
  Choose between a lump sum lottery payment or a series of annual payments.
  Evaluate a personal investment.
• And as an aside, ***the mathematics of engineering economics is trivial***. There are no derivatives, integrals, or differential equations. A few simple algebraic equations and sums are used, but the range of engineering and personal finance problems that they can solve is *limitless*.

1.2 TEXT OBJECTIVES

The purpose of this text is to show you how to incorporate engineering economics into the following *engineering* tasks:

1. ***Comparing two or more cost alternatives***, such as different structural shapes (civil engineering), reactors (chemical engineering), air conditioning systems (mechanical engineering), and electric generators (electrical engineering). This is covered in **Chapters 4 and 5**.
2. ***Calculating the rate of return of a simple investment***, such as an office building (civil engineering), a chemical plant (chemical engineering), a bottling plant (mechanical engineering), and a power plant (electrical engineering). This is covered in **Chapter 6**.
3. ***Calculating the rate of return of multi-simple & non-simple investments*** is explained and illustrated in **Chapter 7**.
4. ***Selecting the best simple investment from two or more mutually exclusive investments*** *(only one can be chosen)*, such as two or more competing office building designs (civil engineering), chemical plant designs (chemical engineering), bottling plant designs (mechanical engineering), and power plant designs (electrical engineering). This is covered in **Chapter 8**.
5. ***Incorporating state and federal income taxes into a rate of return calculation.*** This is covered in **Chapter 9**.
6. ***Determining the sensitivity of the rate of return to a change in the value of an economic variable***, such as the selling price of a product or the fixed capital required to build a manufacturing plant. This is covered in **Chapter 10**.
7. ***Using profitability measures that are different from the rate of return.*** This is covered in **Chapter 11**.

To carry out these engineering tasks, you need to understand the foundation of engineering economics. This is covered in **Chapters 2** and **3**.

In addition to the technical applications of engineering economics, the book presents numerous examples and applications of it that can be used in your personal life. For example, when you graduate you may purchase a house and finance it with a mortgage. There is an Excel spreadsheet in **Chapter 3** that calculates the yearly interest of a mortgage for arbitrary values of the size of the loan and its term and interest rate. You will see that *the total interest paid can be more than the size of the loan itself!*

There are several examples that illustrate the investment cash flows required to fund different retirement cash flow scenarios. And if you win the lottery, you will be able to compare the economic worths of a lump sum payment and a uniform series of 10 or 20 annual payments.

2

Foundation of Engineering Economics

OBJECTIVES

Loans and investments differ only in their terminologies and levels of risk. The mathematical model for calculating the rate of return of an investment is identical to that for calculating an unknown interest rate of a loan. Hence, ***loan concepts are the foundation of engineering economics***. Chapter 2 explains them and illustrates their use in solving *four* major classes of problems in engineering economics. After reading this chapter, you will be able to

1. Represent the transfers of money in a loan or investment with a *cash flow diagram*.
2. Use the ***12-FACTOR Calculator*** spreadsheet to obtain the values of the 12 factors for *arbitrary* values of $i\%$, N, and $g\% \neq i\%$.
3. Use one, two, or all of the three *basic series* in calculations.
4. Make various *equivalence* calculations, ***E*-Class** problems.
5. Make *worth equality* calculations, ***WE*-Class** problems.
6. Perform *present worth comparison* calculations, ***PWC*-Class** problems.
7. Calculate the *rate of return* (ROR) of a simple investment, an ***SI*-Class** problem, with Excel's IRR function and modify one of the cash flows with a Goal Seek iteration to obtain a desired ROR.

2.1 LOAN ELEMENTS

If you have credit card debt or some other type of loan, such as a car loan, student loan, or even a home mortgage, you probably are familiar with some of the terms used in loan calculations. Section 2.1 defines them all.

2.1.1 Monies Transferred

A **simple consumer loan** transfers an amount of money (the **principle** P) *from* a lender *to* a borrower subject to the **condition** that the borrower will transfer N *equal* amounts of money (the **payments** $R_1, R_2, \ldots, R_N$) *to* the lender over a *specified* time interval called the **term** T. Some typical terms are 2 years (furniture loan), 3 years (automobile lease), and 30 years (home mortgage). The payments return the principle and the **total interest** I_T earned by the loan for the use of the borrowed money.

2.1.2 Balances, Interest Rate & Total Interest

Number of Interest Periods & Loan Timeline

To calculate the total interest earned by a loan, the lender divides its term into one or more *equal* time intervals called **interest periods** p. *Interest is earned in each of them.* The **number of interest periods** N is just T divided by p:

$$N = \frac{T}{p}, \tag{2-1}$$

where the units of p and T must be the same, that is, ***N* is dimensionless**. For example, if the term is four years and the interest period is one month,

$$N = \frac{T}{p} = \frac{4\,\text{a} \times 12\,\text{mo·a}^{-1}}{1\,\text{mo}} = 48,$$

where "a" and "mo" are the symbols for year (annual) and month, respectively. The **customary interest periods** are 1 d, 1 w, 1 mo, 3 mo, 6 mo, and 1 a, where "d" and "w" are the symbols for day and week, respectively.

Figure 2.1 is the **timeline** of a loan. It depicts the N interest periods in the term. The **present time** PT occurs at the *beginning* of the *first* interest period, and the **future time** FT occurs at the *end* of the Nth interest period. The numbers under the timeline ticks are called **timeline points**, and a *specific* one is denoted by TP_k where $k = 0, 1, ..., N$.

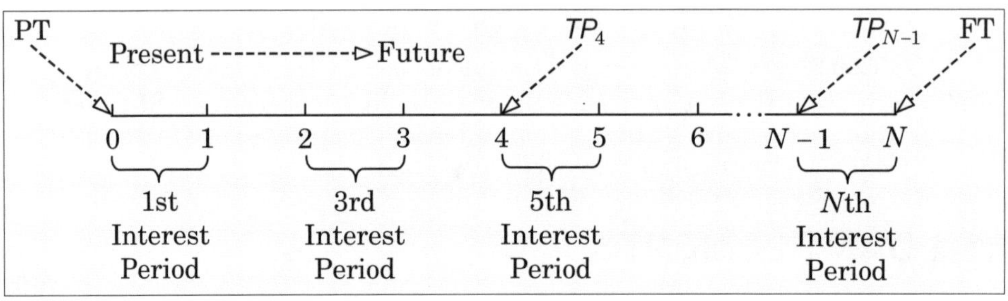

Figure 2.1 **Timeline of a Loan** Note: An interest period is $\frac{1}{N}$th of the term

Balances & Interest Rate

The money owed at the beginning or end of an interest period is called a **balance** B, and the balance at the *end* of the kth interest period is denoted by B_k. Figure 2.1 shows that the balance at the *beginning* of an interest period is the balance at the *end* of the *preceding* interest period. That is, the balance at the beginning of the kth interest period is B_{k-1}.

The **interest rate** i, which is occasionally called the **periodic interest rate**, is *defined* as the ratio of the **periodic interest** I_k earned *in* the kth interest period and the balance at the *beginning* of the kth interest period B_{k-1}:

$$i \equiv \frac{I_k}{B_{k-1}} \qquad k = 1, 2, \ldots, N, \tag{2-2}$$

where the symbol ≡ means "is defined as."

The interest rate is a dimensionless ratio that specifies the interest earned by a unit balance. *The lender normally specifies the value of i.* (See the Practice Note after Example 2.2 for the exception to this rule.) For virtually all loans, the value of i is a small decimal fraction, such as 0.0004932, 0.0075, or 0.06.

Practice Note: *The interest rate is normally constant during the term. Two notable exceptions are the interest rates of adjustable rate mortgages and some student loans.*

Total Interest

The **total interest** I_T earned by a loan is the sum of the interests earned in its N interest periods:

$$I_T = \sum_{k=1}^{N} I_k. \tag{2-3}$$

A convenient way to calculate I_T is to recognize that it must equal the **total payment** R_T minus the **total principle** P_T (see Subsection 2.5.3 for the proof):

$$\begin{aligned} I_T &= R_T - P_T \\ &= \sum_{k=0}^{N} R_k - \sum_{k=0}^{N} P_k, \end{aligned} \tag{2-4}$$

where k is *any* point on the loan timeline. NOTES: (1) P_0 and R_0 are transferred at the *beginning* of the *first* interest period. (2) Every other R_k or P_k is a payment received or principle disbursed at the *end* of the kth interest period.

2.1.3 Interest Rate Specifications

It is customary to include the value of i and the unit of p in an **interest rate specification**. This is often done by placing "per" and the specified unit of p after the value of i. For example, if $i = 0.000493151$ and $p = 1$ d, any of the following interest rate specifications can be used to *simultaneously* specify the value of i and the unit of p:

- $i = 0.000493151$ per day
- the interest rate is 0.000493151 per day
- the daily interest rate is 0.000493151

Percentage Interest Rate

Since the value of i is normally a cumbersome decimal fraction, it is convenient to *define* the **percentage interest rate** $i\%$ as

$$i\% \equiv 100i. \tag{2-5}$$

For example, if $i = 0.01$ per month, we say that the interest rate is 1% per month.

Annual Percentage Rate

The percentage interest rate is useful when $i \geq 0.01$. When this is not so, it usually means that there are many interest periods in a year. To simplify the specification of an interest rate when $i \leq 0.01$, an **annual percentage rate** *APR* is *defined* as

$$APR \equiv 100mi = 100r, \tag{2-6}$$

where m, which is dimensionless, is the **number of interest periods in a year**:

$$m \equiv \frac{1\text{a}}{p}. \tag{2-7}$$

The product mi in Equation (2-6) is called the **nominal interest rate** r. The values of m for the customary interest periods are shown below:

INTEREST PERIOD p	1 d	1 w	1 mo	3 mo	6 mo	1 a
VALUE OF m	365	52	12	4	2	1

Let's use Equation (2-6) to calculate the *APR* for a typical credit card interest rate of 0.000493151 per day:

$$APR = 100\frac{1\text{a}}{p}i = 100\frac{1\text{a}\times 365\ \text{d}\cdot\text{a}^{-1}}{1\text{d}}0.000493151 = 18\%\ \text{compounded daily.}$$

Usage Note: *The specification of an APR consists of a number followed by "compounded" and an adverb that indicates the size of the interest period: daily, weekly, monthly, quarterly, semiannually, or annually.*

To convert an *APR* to an interest rate, we solve Equation (2-6) for i:

$$i = \frac{APR}{100m}. \tag{2-8}$$

For example, if the *APR* is 9% compounded monthly, $m = 12$, and

$$\begin{aligned} i &= \frac{APR}{100\times m} \\ &= \frac{9}{100\times 12} = 0.00705 \text{ per month.} \end{aligned}$$

⚠ ***Regardless of how the interest rate is stated, i must be used in the calculations of engineering economics!***

2.1.4 Analogy Between Loans & Rentals

Borrowing money is like renting it. Table 2.1 compares the terminologies for the rentals of money when interest is paid when earned and an apartment.

TABLE 2.1 Terminologies of Money and Apartment Rentals

RENTAL ITEM	TERMINOLOGY	
	MONEY	APARTMENT
Person Lending the Object	Lender	Landlord
Person Receiving the Object	Borrower	Tenant
Length of Rental	Term	Term
Money Paid for the Rental	Interest	Rent
Payment Time	End of Interest Period	End of Month
End of Term Event	Return of Principle	Return of Apartment

2.2 CASH FLOWS

2.2.1 Definitions

Money is transferred—received or disbursed—during the term. Each transfer is an algebraic quantity called a **cash flow**. Its **magnitude** is the *amount* of money transferred; *it is always a positive number*. The **sign** of a cash flow depends on the direction of the flow. From the lender's point-of-view, a payment cash flow is positive, and a principle cash flow is negative. From the borrower's point-of-view, the opposite is true. A group of related cash flows, such as principles or payments, is called a **series**.

Book Convention: *The lender's point-of-view is normally used to describe cash flows.*

2.2.2 Cash Flow Diagram

A **cash flow diagram** (CFD) is a graphical representation of the cash flows that occur on the timeline of a loan, an investment, a series, or a problem in engineering economics. *Cash flows are represented by vertical arrows* placed at each timeline point where a cash flow occurs. Figure 2.2 is the CFD of a loan, with one principle at TP_0, when interest is paid when earned. The payments for the first $N-1$ interest periods are each Pi. Since *the last payment at TP_N always reduces the balance to zero*, it is the sum of the interest earned in the Nth interest period and the principle.

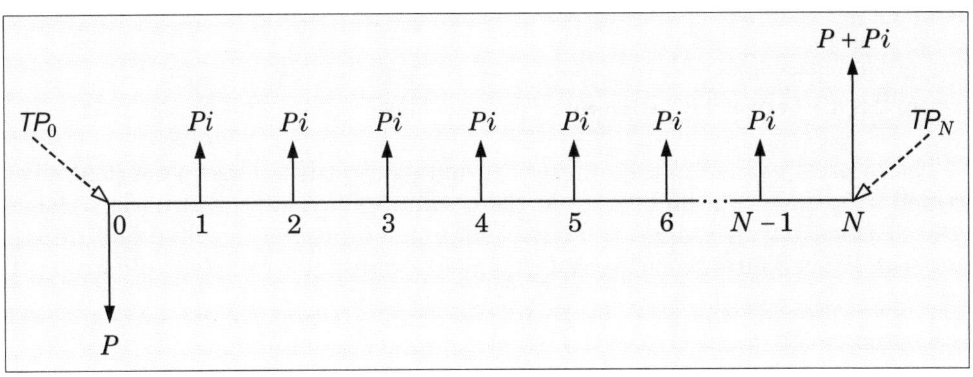

Figure 2.2 **CFD of a Loan** NOTE: This CFD is drawn from the point-of-view of the lender.

Practice Note: *Here are the* ***Arrow Drawing and Labeling Conventions*** *for the directions, labels, and sizes of the arrows* ***of a CFD****:*

Arrow Drawing and Labeling Conventions of a CFD
(1) The arrows of positive cash flows point up, and those of negative cash flows point down.
(2) Depending on the direction of an arrow, its label is placed above or below the arrowhead. It may be the *magnitude* of the cash flow ($500) or a letter ($P$), subscripted letter ($R_N$), or simple expression ($P+Pi$) that represents it.
(3A) Arrows are not drawn to scale. Their lengths are roughly proportional to the magnitudes of their cash flows.
(3B) Arrows of cash flows with equal magnitudes should have equal lengths.

2.3 FUNDAMENTAL REPAYMENT EQUATION

2.3.1 Definitions

The **repayment equation** of a loan is the relation between its principles, payments, interest rate i, and number of interest periods N. The **Fundamental Repayment Equation** (FRE) is based on the **one principle-one payment** ($1P$-$1R$) loan. Its CFD is displayed in Figure 2.3. A single principle is always denoted by P, and the payment is denoted by R_N because it occurs at the end of the Nth interest period.

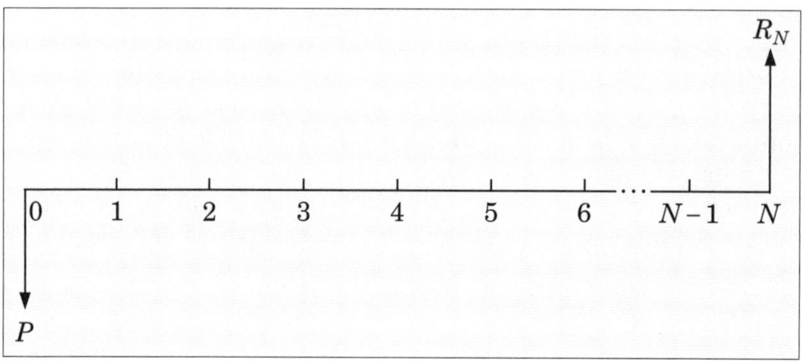

Figure 2.3 **CFD of a 1*P*-1*R* Loan**

The interest earned in any interest period—except the last—is not paid when earned. It is *added* to the balance at the *beginning* of the interest period. Hence, *interest is earned on unpaid interest*. This type of interest is called **compound interest**. When interest is not earned on unpaid interest, it is called **simple interest**.

The interest-earning capacities of compound and simple interests are compared in Subsection 3.2.2.

Practice Note: *The overwhelming majority of loans are based on compound interest.*

Book Convention: *Unless stated otherwise, "interest" means compound interest.*

2.3.2 Derivation

Since interest is not paid when earned, the balance at the end of the first interest period is the sum of the money owed at the beginning of the first interest period and the interest earned in it:

$$\begin{aligned} B_1 &= P + I_1 = P + Pi \\ &= P(1+i). \end{aligned} \tag{2-9}$$

The balance at the end of the second interest period is obtained in the same way:

$$B_2 = B_1 + I_2 = B_1(1+i).$$

Substituting B_1 from Equation (2-9) into this equation yields

$$B_2 = P(1+i)^2. \tag{2-10}$$

Equations (2-9) and (2-10) *suggest* that the balance at the end of the kth interest period has the same form:

$$B_k = P(1+i)^k. \tag{2-11}$$

If Equation (2-11) is valid,

$$\begin{aligned} B_{k+1} &= B_k(1+i) = P(1+i)^k(1+i) \\ &= P(1+i)^{k+1}. \end{aligned} \tag{2-12}$$

Since Equation (2-11) is valid for $k = 1$, it follows from Equation (2-12)that Equation (2-11) is valid for $k = 1, 2, \ldots, N-1$.

The proof of Equation (2-11) is an example of a proof by **mathematical induction**. It consists of two propositions:

- *Proposition 1:* Prove that Equation (2-11) is valid for $k = 1$.
- *Proposition 2:* Assume the validity of Equation (2-11) and show that Equation (2-12) has the same form as Equation (2-11).

If both propositions are true, Equation (2-11) is valid for $k = 1, 2, \ldots, N-1$.

At the end of the term, the payment must reduce the balance to 0. Hence,

$$\begin{aligned} B_N &= B_{N-1} + I_N - R_N \\ 0 &= P(1+i)^{N-1}(1+i) - R_N \\ 0 &= P(1+i)^N - R_N. \end{aligned}$$

Solving this equation for the payment yields

$$R_N = P(1+i)^N. \tag{2-13}$$

It is customary in the literature of engineering economics to denote an amount of money in the future by F. Hence, the FRE becomes

$$\boxed{F = P(1+i)^N.} \tag{2-14}$$

Computation Note: *The FRE contains four variables: F, P, N, and i. To solve it for F, P, or N, we must specify the values of the other three variables. This type of problem is an example of an* ***E-Class*** *problem. Its solution is based on the concept of* ***equivalence****, which is defined in Subsection 2.4.5.*

I call Equation (2-14) the Fundamental Repayment Equation because it is the basis of the *definitions* of the future and present worths of a cash flow. These worths are the basis of the concept of ***"the time value of money,"*** and they are used to *define* the present and future worths of *any* series.

2.3.3 Total & Normalized Total Interests

We use Equation (2-4) to calculate the total interest of *any* loan:

$$I_T = R_T - P_T. \tag{2-4}$$

Hence, it follows from Equation (2-14) that the **total interest** I_T of a 1P-1R loan is given by

$$I_T = F - P = P\left[(1+i)^N - 1\right]. \tag{2-15}$$

The **normalized total interest** $I_T P^{-1}$ only depends on i and N:

$$\frac{I_T}{P} = (1+i)^N - 1. \tag{2-16}$$

2.3.4 Using the Fundamental Repayment Equation

EXAMPLE 2.1 A Loan That Almost Doubles the Principle in 10 Years: Calculating *F* (*E*-Class)

Problem: A single payment loan has a principle of \$1,000. The term is 10 years and the interest rate is 7% per year. *How much money does the lender receive at the end of the 10th year, and how much interest has the loan earned?*

Solution: Here the unknown is F. The principle and term are specified: $P = \$1{,}000$, and $T = 10\,\text{a}$. To calculate F, we must determine the values of i and N.

Since the interest rate is 7% per year, $i = 0.07$, and $p = 1\,\text{a}$ (see Equation (2-5)). We use Equation (2-1) to calculate N:

$$N = \frac{T}{p} = \frac{10\,\text{a}}{1\,\text{a}} = 10.$$

To calculate F, we use Equation (2-14):

$$F = P(1+i)^N = \$1{,}000(1+0.07)^{10} = \underline{\underline{\$1{,}967.15}}.$$

Equation (2-16) is used to calculate I_{T}:

$$\begin{aligned} I_{\mathrm{T}} &= F - P \\ &= \$1{,}967.15 - \$1{,}000 = \underline{\underline{\$967.15}}. \end{aligned}$$

Remark: *The preceding result shows that an amount of money loaned at 7% per interest period is nearly doubled in 10 interest periods.*

EXAMPLE 2.2 Saving for the Future: Calculating *P* (*E*-Class)

Problem: *How much money must be deposited in a bank account to provide \$10,000 in three years if the APR is 6% compounded daily?*

Solution: Here the unknown is P. The future amount of money and the term are specified: $F = \$10{,}000$, and $T = 3\,\mathrm{a}$. To calculate P, we must determine the values of i and N.

It follows from the *APR* that $p = 1\,\mathrm{d}$. Hence, there are 365 interest periods per year; that is, $m = 365$. We use Equation (2-8) to calculate i:

$$i = \frac{APR}{100m} = \frac{6}{100 \times 365} = 0.000164384 \text{ per day}.$$

Equation (2-1) is used to calculate N:

$$N = \frac{T}{p} = \frac{3\,\mathrm{a} \times 365\,\mathrm{d{\cdot}a^{-1}}}{1\mathrm{d}} = 1{,}095.$$

We rearrange Equation (2-14) to calculate P:

$$P = \frac{F}{(1+i)^N} = \frac{\$10{,}000}{(1+0.000164384)^{1{,}095}} = \underline{\underline{\$8{,}352.82}}.$$

Practice Note: *A savings account is a loan that a depositor makes* ***to*** *a bank. The depositor specifies the principles (deposits), payments (withdrawals), their timing, and the term, but the* ***bank (borrower) specifies the interest rate!***

EXAMPLE 2.3 Time to Grow Money: Calculating *N* (*E*-Class)

Problem: How many days are required to grow \$5,000 to \$10,000 in a bank account that has an *APR* of 9% compounded daily?

Solution: Here the unknown is N. The principle and the future amount of money are specified: $P = \$5{,}000$, and $F = \$10{,}000$. To calculate N, we must determine the value of i.

It follows from the *APR* that $p = 1\,\text{d}$. Hence, $m = 365$. Equation (2-8) is used to calculate i:

$$i = \frac{APR}{100m} = \frac{9}{100 \times 365} = 0.000246575 \text{ per day.}$$

To calculate N, we first divide Equation (2-14) by P:

$$\frac{F}{P} = (1+i)^N. \tag{E2.3-1}$$

The natural logarithms of the sides of Equation (E2.3-1) are equal:

$$\ln\frac{F}{P} = N\ln(1+i). \tag{E2.3-2}$$

Dividing both sides of Equation (E2.3-2) by $\ln(1+i)$ yields the equation for calculating N:

$$N = \frac{\ln\frac{F}{P}}{\ln(1+i)} = \frac{\ln\frac{\$10{,}000}{\$5{,}000}}{\ln(1+0.000246575)} = \underline{\underline{2{,}811.45}}. \tag{E2.3-3}$$

EXAMPLE 2.4 Calculating *F* and the Total Interest of a Loan with Two Interest Rates (*E*-Class)

Problem: A customer deposits \$15,000 in a savings account whose interest rate is 5% compounded daily. Two years later it decreases to 3% compounded daily and remains so for the next three years. *(1) How much money can be withdrawn at the end of five years? (2) What is the total interest?*

Solution: We divide the loan to the bank into two consecutive loans with different terms and interest rates. The principle of the second loan is the balance at the end of the first loan.

(1) Final Balance of the First Loan

The principle and term are specified: $P_1 = \$15{,}000$, and $T_1 = 2\,\text{a}$. To calculate the balance at the end of the term, we must determine the values of i_1 and N_1.

It follows from APR_1 that $p_1 = 1\,\text{d}$. Hence, $m_1 = 365$. Equation (2-8) is used to calculate i_1:

$$i_1 = \frac{APR_1}{100m_1} = \frac{5}{100 \times 365} = 0.000136986 \text{ per day.}$$

We calculate N_1 with Equation (2-1):

$$N_1 = \frac{T_1}{p_1} = \frac{2\,\text{a} \times 365\,\text{d}\cdot\text{a}^{-1}}{1\,\text{d}} = 730.$$

Lastly, we use Equation (2-11) to calculate B_{N_1}:

$$B_{N_1} = P_1(1+i_1)^{N_1} = \$15{,}000(1+0.000136986)^{730} = \$16{,}577.45.$$

(2) Total Interest of the Two Loans

The second principle and term are specified: $P_2 = B_{N_1} = \$16{,}577.45$, and $T_2 = 3\,\text{a}$. The second interest rate is given by

$$i_2 = \frac{APR_2}{100m_2} = \frac{3}{100\times 365} = 0.0000821918 \text{ per day.}$$

This calculates the value N_2:

$$N_2 = \frac{T_2}{p_2} = \frac{3\,\text{a}\times 365\,\text{d}\cdot\text{a}^{-1}}{1\,\text{d}} = 1{,}095.$$

Hence, the amount of money that can be withdrawn at the end of the second term, which is the end of the fifth year, is given by Equation (2-14):

$$F = P_2(1+i_2)^{N_2} = \$16{,}577.45(1+0.0000821918)^{1{,}095} = \underline{\underline{\$18{,}138.55}}.$$

Equation (2-4) is used to calculate the total interest earned by the two loans:

$$\begin{aligned} I_{\text{T}} &= F - P_1 \\ &= \$18{,}138.55 - \$15{,}000 = \underline{\underline{\$3{,}138.55}}. \end{aligned}$$

In the previous four examples the interest rates of the loans are specified. However, in the following example the interest rate is the unknown. *Calculating the **interest rate** of a loan or a simple investment is an **SI-Class** problem.*

Syntax Note: *When the interest rate is unknown, we denote it by i^*.*

EXAMPLE 2.5 Calculating the Interest Rate of a Proposed Loan (*SI*-Class)

Problem: A friend wants to borrow \$3,000 for part of the down payment to purchase a camera shop. He *expects* to repay the loan with a single payment of \$6,000 in three years. *What is the anticipated annual interest rate of the proposed loan?*

Solution: The principle, *estimated* payment, *assumed* term, and interest period of this 1*P*-1*R* loan are specified: $P = \$3{,}000$, $F = \$6{,}000$, $T = 3\,\text{a}$, and $p = 1\,\text{a}$. As usual, we calculate N with Equation (2-1):

$$N = \frac{T}{p} = \frac{3\,\text{a}}{1\text{a}} = 3.$$

We calculate i^* as follows: (1) divide Equation (2-14) by P; (2) calculate the Nth root of both sides of the result of Step 1; (3) subtract 1 from both sides of the result of Step 2 to obtain the expression for i^*:

$$i^* = \sqrt[N]{\frac{F}{P}} - 1 = \sqrt[3]{\frac{\$6{,}000}{\$3{,}000}} - 1 = \underline{\underline{0.2599}}. \tag{E2.5-1}$$

Hence, *if all goes as planned for the camera shop*, you would earn 25.99% per year on the proposed $3,000 loan.

Another alternative for you is a $3,000 bank CD earning 6% per year. It will pay you the following *guaranteed*[1] amount F in three years:

$$F = \$3{,}000(1 + 0.06)^3 = \$3{,}573.05.$$

This is $2,426.95 *less* than the payment of the proposed loan. However, *if the camera shop fails, you may not even get your $3,000 back!*

2.4 WORTH, FACTORS & EQUIVALENCE

2.4.1 Worths of a Cash Flow

Time Value of Money

The value of money is called its **worth**, and the **Worth Axiom** governs its change with time:

Worth Axiom
The worth of money increases with time *only* if it is earning interest.

Example 2.1 shows that a present amount of money P is worth approximately $2P$ in 10 years when it earns interest at a rate of 7% per year. However, if the money is stashed under a mattress or buried in a stainless steel box, its worth in 10 years is still P. That is, ***if money is not earning interest, its worth does not increase with time.*** The increase in the worth of money that earns interest over time is the basis of the phrase "**the time value of money.**"

Future Worth

To quantify the increase in the worth of a **present amount of money** P on the timeline when it earns interest at a rate i, we use Equation (2-14) to *define* the **future worth of *P*** (FW_P):

$$\boxed{FW_P \equiv P(1 + i)^N.} \tag{2-17}$$

The **future time** in Equation (2-17) is *defined* as N interest periods *after* the timeline point of P:

[1] Bank accounts are insured up to $250,000 by the Federal Deposit Insurance Corporation.

$$N = \frac{T}{p}, \tag{2-1}$$

where T is the timespan on which interest is earned.

Present Worth

When an amount of money P earns interest, it is worth *more* than P in the future. Hence, a future amount of money F must be worth *less* than F at the present time because its present worth can earn interest and grow to F. The following *definition* of the **present worth of F** (PW_F) *uniquely* satisfies this requirement:

$$\boxed{PW_F \equiv \frac{F}{(1+i)^N}.} \tag{2-18}$$

where i is *identical* to the interest rate used to calculate a future worth, and the **present time** is *defined* as N interest periods *before* the timeline point of F.[2]

Equation (2-19) confirms that the future worth of PW_F is F:

$$\begin{aligned} FW_{PW_F} &= PW_F(1+i)^N \\ &= \frac{F}{(1+i)^N}(1+i)^N = F. \end{aligned} \tag{2-19}$$

Remark: *Calculating the present worth of F is called **discounting** its value.*

EXAMPLE 2.6 Discounting the Future Value of a Bond (*E*-Class)

Problem: An investor owns a corporate bond that will pay $25,000 in three years. However, the company wants to reduce some of its long-term debt by buying this type of bond now. *If the APR of a bond fund for the next three years is expected to be 6% compounded quarterly, what is the minimum amount of money P that the investor should accept to sell the bond now?* NOTE: The interest rate for the problem is the interest rate earned by the bond fund, because the investor will want to invest P in it.

Solution: *To compare the worths of any two amounts of money at different times, we must always compare their worths at a common time.* The amounts for this problem are $25,000 in *three years* and an unknown *equivalent* amount P *now*. To calculate the present worth of the $25,000 future payment, we must first determine the values of i and N. Since the APR is 6% compounded quarterly, $p = 3$ mo and $m = 4$. As usual, we calculate i with Equation (2-8):

$$\begin{aligned} i &= \frac{APR}{100m} \\ &= \frac{6}{100 \times 4} = 0.015 \text{ per quarter.} \end{aligned}$$

The value of N is given by Equation (2-1):

---

[2]Sometimes this present time may not correspond to the present time of a problem.

$$N = \frac{T}{p} = \frac{3\,\text{a} \times 12\,\text{mo}\cdot\text{a}^{-1}}{3\,\text{mo}} = 12.$$

We calculate the present worth of the \$25,000 with Equation (2-18):

$$PW_F = \frac{F}{(1+i)^N} = \frac{\$25{,}000}{(1+0.015)^{12}} = \underline{\underline{\$20{,}909.69}}.$$

If $P \geq PW_F$, the investor can take the offer. NOTE: The decision hinges on the *expected* interest rate of the bond fund over the next three years. If it goes up, the investor got a bargain because when he/she invests P in the bond fund it will grow to more than \$25,000 in three years. But if it goes down, the converse is true.

2.4.2 Factors for a Cash Flow

Definitions: Historical Perspective

Before there were *personal* computers (1980) or electronic calculators (1974) most engineering calculations were *tediously* made with slide rules. Hence, the early practitioners of engineering economics simplified repeated calculations of FW_P and PW_F by tabulating the values of their normalized worths. The **normalized worth of a cash flow** is *defined* as the quotient of its worth and magnitude, which is called its **metric**. Hence, the **normalized future worth of *P*** ($FW_P P^{-1}$) is obtained from Equation (2-17):

$$\frac{FW_P}{P} = (1+i)^N. \tag{2-20}$$

Similarly, the **normalized present worth of *F*** ($PW_F F^{-1}$) follows from Equation (2-18):

$$\frac{PW_F}{F} = \frac{1}{(1+i)^N}. \tag{2-21}$$

The expressions on the right hand sides (RHSs) of these equations are functions of *only* i and N. Hence, their values can be computed and tabulated. To emphasize this, the left hand sides (LHSs) of Equations (2-20) and (2-21) are replaced by **function notations**.[3] However, the early practitioners used a function notation called a **factor**. It has the arguments i and N *and* a mnemonic worth/metric ratio that denotes the normalized worth represented by the factor. The **cash flow future worth factor** $(F/P, i, N)$ is *defined* as

$$\boxed{(F/P, i, N) \equiv \frac{FW_P}{P} = (1+i)^N,} \tag{2-22}$$

and the **cash flow present worth factor** $(P/F, i, N)$ is *defined* as

[3] A function can be denoted by an equation, such as $z = x^2 + 3y$, or with a **function notation**, such as $f(x, y) = x^2 + 3y$.

$$(P/F,i,N) \equiv \frac{PW_F}{F} = \frac{1}{(1+i)^N} = (F/P,i,N)^{-1}. \tag{2-23}$$

Factor Tables

Texts on engineering economics traditionally have a *tabulation* of the values of eight or nine of the 12 factors used in engineering economics. Each page corresponds to a specific *i*% and shows the values of the factors for it and various values of *N*. Figure 2.4 shows part of a page of *tabulated* factor values.

	1.5%								
	Present Worth Factors			**Uniform Worth Factors**			**Future Worth Factors**		
N	***(P/F,i,N)***	***(P/A,i,N)***	***(P/G,i,N)***	***(A/P,i,N)***	***(A/F,i,N)***	***(A/G,i,N)***	***(F/P,i,N)***	***(F/A,i,N)***	***(F/G,i,N)***
1	0.98522	0.98522	0.00000	1.0150	1.0000	0.00000	1.0150	1.0000	0.0000
2	0.97066	1.9559	0.97066	0.51128	0.49628	0.49628	1.0302	2.0150	1.0000
3	0.95632	2.9122	2.8833	0.34338	0.32838	0.99007	1.0457	3.0452	3.0150
4	0.94218	3.8544	5.7098	0.25944	0.24444	1.4814	1.0614	4.0909	6.0602
5	0.92826	4.7826	9.4229	0.20909	0.19409	1.9702	1.0773	5.1523	10.151
6	0.91454	5.6972	13.996	0.17553	0.16053	2.4566	1.0934	6.2296	15.303
7	0.90103	6.5982	19.402	0.15156	0.13656	2.9405	1.1098	7.3230	21.533
8	0.88771	7.4859	25.616	0.13358	0.11858	3.4219	1.1265	8.4328	28.856
9	0.87459	8.3605	32.612	0.11961	0.10461	3.9008	1.1434	9.5593	37.289
10	0.86167	9.2222	40.367	0.10843	0.093434	4.3772	1.1605	10.703	46.848
11	0.84893	10.071	48.857	0.099294	0.084294	4.8512	1.1779	11.863	57.551
12	0.83639	10.908	58.057	0.091680	0.076680	5.3227	1.1956	13.041	69.414
13	0.82403	11.732	67.945	0.085240	0.070240	5.7917	1.2136	14.237	82.455
14	0.81185	12.543	78.499	0.079723	0.064723	6.2582	1.2318	15.450	96.692
15	0.79985	13.343	89.697	0.074944	0.059944	6.7223	1.2502	16.682	112.14

Figure 2.4 **Part of a Page of Tabulated Factor Values** NOTE: The values are normally displayed with five significant digits.

Calculating with Factors

Using factors to calculate a worth of a cash flow is very easy. We calculate FW_P with the value of $(F/P,i,N)$ in a factor table:

$$FW_P = P(F/P,i,N). \tag{2-24}$$

Similarly, we calculate PW_F with the value of $(P/F,i,N)$:

$$PW_F = F(P/F,i,N). \tag{2-25}$$

Let's use Equation (2-25) and Figure 2.4 to recalculate the present worth of the bond in Example 2.6. Since $F = \$25{,}000$, $i\% = 1.5\%$ per quarter, and $N = 12$,

$$PW_F = \$25{,}000\underbrace{(P/F,1.5\%,12)}_{0.83639} = \$20{,}910.$$

This result is identical to the result in Example 2.6 when it is rounded to the nearest dollar. As you can see, the mnemonic function notation is simple, and using the values of *i%* and *N* in the factor is self-explanatory.

Usage Notes: *Using factors in calculations has three advantages over using algebraic worth expressions:*
(1) You don't have to remember or look up algebraic worth expressions.
(2) An equation with factors is much simpler to write, read, and evaluate than one with algebraic worth expressions.
(3) The interest rate in a factor can be displayed as i% rather than the cumbersome interest rate i required for an evaluation of an algebraic worth expression.

Book Convention: *A factor value is always placed beneath the factor. This makes it easy to follow the calculation in which the factor is used.*

2.4.3 *12-FACTOR Calculator*

Limitations of Factor Tables

Typical tabulations of factor values are useful, but they have inherent limitations:

1. Factor values for *arbitrary* interest rates *i%* can't be tabulated.
2. Factor values for *arbitrary* values of *N* can't be tabulated.
3. The number of significant digits of a factor value can't be changed.

In addition, **the present, future, and uniform worth factors of a *geometric gradient series*, which is explained in Section 2.8, can't be conveniently tabulated because they depend on *three* variables: *i%*, *g%*, and *N*.**

Creating the 12-FACTOR Calculator

To eliminate the preceding limitations, we use the ***12-FACTOR Calculator***, displayed in Spreadsheet 2.1 on the next page. It's an Excel spreadsheet created by the author and located in the *Book Templates* file provided with this book. It calculates the 12 factors of engineering economics *for **arbitrary** values of i%, N, and g% ≠ i%.*

Section 2.12 explains the development of the ***12-FACTOR Calculator***.

Using the 12-FACTOR Calculator

You just enter values for *i%*, *six* values of *N* (occasionally you need more than one value of *N* in a problem), and *g% ≠ i%*, and the ***12-FACTOR Calculator*** *immediately* displays the values of the 12 factors. If you *accidentally* enter the same values for *i%* and *g%*, Excel returns error messages (#VALUE!) for the 18 factor values of the geometric gradient series. If you leave out an *N* value, Excel assumes its 0 and returns error messages (#VALUE!) for the corresponding four uniform worth factor values.

Usage Note: Every cell except the eight input cells is **protected** to prevent the accidental deletion or modification of its contents. However, to increase the number of displayed decimal places of a cell beyond the default value of five, *Right Click* the cell > *Click* **Format Cells...** > *Select* the desired value in the **Decimal places:** list box > *Click* **OK**. If there is not enough space in the cell to display the number, Excel returns ######. You can increase the column width by *Clicking* on the column heading and **dragging its right border** to the right to increase the cell width.

	A	B	C	D	E	F	G	H
1								
2		***12-FACTOR Calculator***						
3				***i*%**	***g*%**			
4			***Input:***	1.5%	4%			
5			***Number of Interest Periods N***					
6		***Input:***	2	4	6	8	10	12
7			***Solution: Present Worth Factor Values***					
8		$(P/F, i, N)$	0.97066	0.94218	0.91454	0.88771	0.86167	0.83639
9		$(P/A, i, N)$	1.95588	3.85438	5.69719	7.48593	9.22218	10.90751
10		$(P/G, i, N)$	0.97066	5.70985	13.99560	25.61574	40.36748	58.05708
11		$(P/A_1, g, i, N)$	1.99471	4.08889	6.28751	8.59576	11.01912	13.56333
12			***Solution: Annual Worth Factor Values***					
13		$(A/P, i, N)$	0.51128	0.25944	0.17553	0.13358	0.10843	0.09168
14		$(A/F, i, N)$	0.496278	0.24444	0.16053	0.11858	0.09343	0.07668
15		$(A/G, i, N)$	0.49628	1.48139	2.45658	3.42185	4.37721	5.32267
16		$(A/A_1, g, i, N)$	1.01985	1.06084	1.10362	1.14826	1.19485	1.24349
17			***Solution: Future Worth Factor Values***					
18		$(F/P, i, N)$	1.03023	1.06136	1.09344	1.12649	1.16054	1.19562
19		$(F/A, i, N)$	2.01500	4.09090	6.22955	8.43284	10.70272	13.04121
20		$(F/G, i, N)$	1.00000	6.06022	15.30340	28.85594	46.84811	69.41410
21		$(F/A_1, g, i, N)$	2.05500	4.33980	6.87503	9.68306	12.78814	16.21656

Spreadsheet 2.1 ***12-FACTOR Calculator*** NOTES: (1) Every factor value used in this book was obtained with the ***12-FACTOR Calculator***. To use it, open and create a copy of the *Book Templates* file (It's located at the Wiley Internet site for this book.) and click on the ***12-FACTOR Calculator*** tab. (2) *The factor values here are identical to the corresponding* ***tabulated*** *values in Figure 2.4.*

2.4.4 Present & Future Worths of a Series

Figure 2.5 shows a CFD of an *arbitrary* series called X. We *define* its **present worth** PW_X as the sum of the present worths of its cash flows at its **series present time** SPT, which is *defined* as the location of its first cash X_0. The present worth of the series is calculated on the following page:

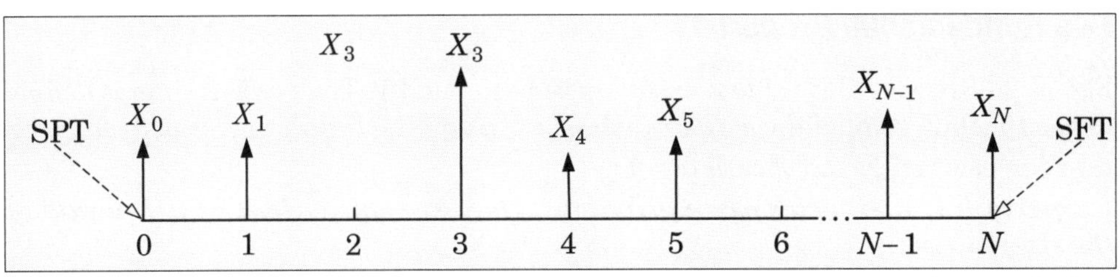

Figure 2.5 **Cash Flows of the *X* Series** NOTE: The *location* of its *first* cash flow defines its TP_0, and the *location* of its *last* cash flow defines its TP_N.

$$\begin{aligned} PW_X &\equiv X_0 + \frac{X_1}{(1+i)} + \frac{X_2}{(1+i)^2} + \cdots + \frac{X_N}{(1+i)^N} \\ &\equiv \sum_{k=0}^{N} \frac{X_k}{(1+i)^k}. \end{aligned} \tag{2-26}$$

We *define* the **worth** $W_{X,k}$ **of the *X* series** at *any* TP_k as

$$\begin{aligned} W_{X,k} &\equiv X_0(1+i)^{k-0} + X_1(1+i)^{k-1} + \cdots + X_k(1+i)^{k-k} + \cdots + X_N(1+i)^{k-N} \\ &= (1+i)^k \sum_{k=0}^{N} \frac{X_k}{(1+i)^k}. \end{aligned} \tag{2-27}$$

Hence, it follows from Equation (2-26) that

$$\boxed{W_{X,k} = PW_X(1+i)^k.} \tag{2-28}$$

***Usage Note**: Once we obtain the present worth of a series at its SPT, we use Equation (2-28) to calculate its worth at any timeline point k interest periods greater or less than its SPT.*

The **future worth of the *X* series** FW_X is *defined* as its worth at the **series future time** N_X. Hence, it follows from Equation (2-28) that

$$FW_X = PW_X(1+i)^{N_X}. \tag{2-29}$$

2.4.5 Equivalence

Definition

Two cash flows, two series, or a series and a cash flow are ***equivalent*** *if, for the same interest rate* i_Q, *they have equal worths at one point on the timeline.* For example, the FRE shows that the present worths of the payment F and principle P of the ($1P$-$1R$) loan are equal. Hence, ***the payment and principle are equivalent.***

If the payment and principle were not equivalent, it would be unfair to the lender or the borrower. For example, if the present worth of the payment F exceeds the principle P, the lender receives more money than he has loaned when the worths of the payment and principle are measured by their present worths at the interest rate of the loan.

Two Fundamental Properties

Figure 2.6 shows a CFD of two *arbitrary* series U and V. The present time is *defined* as the timeline point of the *earliest* cash flow V_0, and the future time is *defined* as the timeline point of the *latest* cash flow U_8.

Property 1: If two series are equivalent, they are equivalent at all points on the timeline.

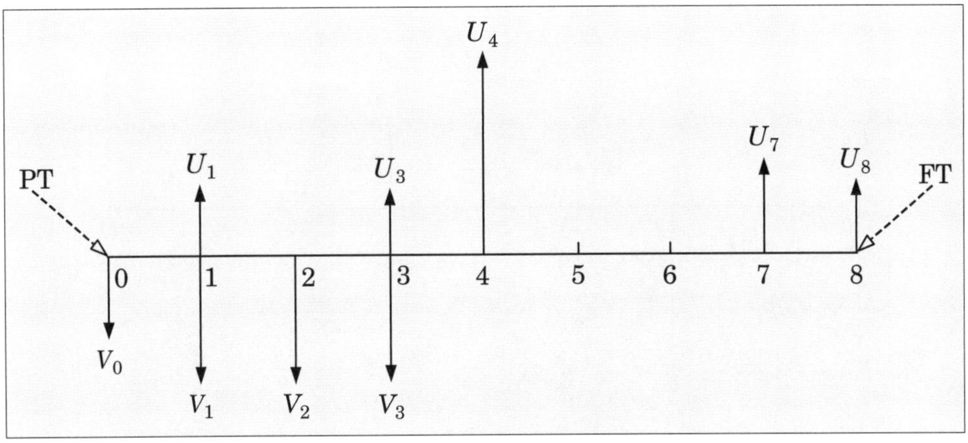

Figure 2.6 **CFD of Two Arbitrary Series U and V**

Proof: If series U and V are equivalent at a *particular* TP_k, it follows from Equation (2-28) that

$$PW_U = PW_V.$$

Hence, it also follows from Equation (2-28) that

$$\boxed{W_{U,k} = W_{V,k},}$$

where k is *any* point on the timeline. That is, *the worths of equivalent series U and V are equal at all points on the timeline.* Q.E.D.

Property 2: The equivalence interest rate i_Q is UNIQUE when there is only ONE change in the signs of the INCREMENTAL CASH FLOWS, $CF_k \equiv U_k - V_k$. Changing the interest rate destroys the equivalence!

Proof: Subsection 2.9.3 shows that when there is only one change in the signs of the cash flows of any series, there is only one interest rate at which the present worth of the series equals zero. Hence, the series U and V are equivalent at that interest rate. That is, they have the same worths at that interest rate.

2.5 GENERAL REPAYMENT EQUATION (GRE)

The **General Repayment Equation** (GRE) is the relationship between the principles, payments, interest rate i, and number of interest periods N of *any* loan, such as the one in Figure 2.7. NOTE: ***A loan is always structured so that every balance except B_n is positive.*** To derive the GRE, we need to define the Balance Equation.

2.5.1 Balance Equation

Only earned interest causes the balance of a $1P$-$1R$ loan to change:

$$B_k = B_{k-1} + I_k \qquad k = 1, 2, \ldots, N. \tag{2-30}$$

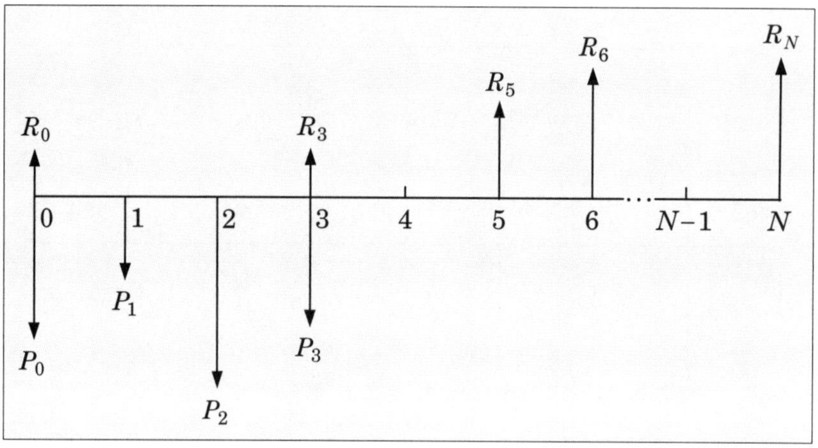

Figure 2.7 **CFD of a Loan with Irregular Principle and Payment Series**
NOTES: (1) *A payment and principle, such as R_3 and P_3, can occur at one or more timeline points.* (2) *A cash flow does not have to occur at every point on the timeline:* there are no cash flows at the ends of the 4th and $(N-1)$th interest periods.

However, as shown in Figure 2.7, at the end of the of kth interest period, a principle may be disbursed *and* a payment may be received. Hence, adding $P_k - R_k$ to Equation (2-30) yields the **Balance Equation**:

$$\boxed{B_k = B_{k-1}(1+i) + P_k - R_k \qquad k = 1, 2, \dots, N,} \tag{2-31}$$

where the interest earned in the kth interest period I_k is just

$$I_k = B_{k-1} i \qquad k = 1, 2, \dots, N. \tag{2-32}$$

Normally, B_0 is just P_0. However, a lender may occasionally require an initial payment R_0 as an inducement to make a loan. Hence, B_0—*which like all balances except B_N is always positive*—is the difference of P_0 and R_0:

$$B_0 = P_0 - R_0 > 0. \tag{2-33}$$

At the end of the term the balance must be 0:

$$B_N \equiv 0. \tag{2-34}$$

2.5.2 Derivation of the GRE

To derive the GRE, we first rearrange Equation (2-31) so that only the balance terms are on one side:

$$R_k - P_k = B_{k-1}(1+i) - B_k \qquad k = 1, 2, \dots, N.$$

Next, we multiply both sides of this equation by $(1+i)^{-k}$:

$$\frac{R_k}{(1+i)^k} - \frac{P_k}{(1+i)^k} = \frac{B_{k-1}}{(1+i)^{k-1}} - \frac{B_k}{(1+i)^k} \qquad k = 1, 2, \ldots, N. \tag{2-35}$$

When we sum both sides of Equation (2-35), the RHS (right hand side) reduces to two terms because pairs of identical terms cancel each other:

$$\begin{aligned}
\sum_{k=1}^{N} \frac{R_k}{(1+i)^k} - \sum_{k=1}^{N} \frac{P_k}{(1+i)^k} &= \sum_{k=1}^{N} \left[\frac{B_{k-1}}{(1+i)^{k-1}} - \frac{B_k}{(1+i)^k} \right] \\
&= \left[\frac{B_0}{(1+i)^0} - \frac{\cancel{B_1}}{\cancel{(1+i)^1}} \right] + \left[\frac{\cancel{B_1}}{\cancel{(1+i)^1}} - \frac{\cancel{B_2}}{\cancel{(1+i)^2}} \right] + \cdots \\
&\quad + \left[\frac{\cancel{B_{N-2}}}{\cancel{(1+i)^{N-2}}} - \frac{\cancel{B_{N-1}}}{\cancel{(1+i)^{N-1}}} \right] + \left[\frac{\cancel{B_{N-1}}}{\cancel{(1+i)^{N-1}}} - \frac{B_N}{(1+i)^N} \right] \\
&= \frac{B_0}{(1+i)^0} - \frac{B_N}{(1+i)^N}.
\end{aligned} \tag{2-36}$$

Since $B_N \equiv 0$ and $B_0 = P_0 - R_0$, Equation (2-36) reduces to the GRE:

$$\boxed{\sum_{k=0}^{N} \frac{R_k}{(1+i)^k} = \sum_{k=0}^{N} \frac{P_k}{(1+i)^k}.} \tag{2-37}$$

Hence, ***the present worth of the payment series equals the present worth of the principle series, i.e., the principle and payment series are EQUIVALENT!***

Computation Note: *The GRE plays a central role in engineering economics because it leads naturally and directly to the definition of the* ***internal rate of return*** *(IRR)—the unique interest rate of a simple investment (see Subsection 2.9.3).*

2.5.3 Total Interest Equation

We now derive Equation (2-4), the total interest equation. It follows from Equation (2-31) that we can write the Balance Equation as

$$B_k - B_{k-1} = I_k + P_k - R_k \qquad k = 1, 2, \ldots, N. \tag{2-38}$$

Summing both sides of Equation (2-38) yields

$$\begin{aligned}
\sum_{k=1}^{N} B_k - \sum_{k=1}^{N} B_{k-1} &= \sum_{k=1}^{N} I_k + \sum_{k=1}^{N} P_k - \sum_{k=1}^{N} R_k \\
B_N - B_0 &= I_T + \sum_{k=1}^{N} P_k - \sum_{k=1}^{N} R_k.
\end{aligned}$$

Since $B_N \equiv 0$ and $B_0 = P_0 - R_0$, this equation reduces to Equation (2-4):

$$I_T = R_T - P_T, \tag{2-4}$$

where

$$P_{\mathrm{T}} \equiv \sum_{k=0}^{N} P_k$$
$$R_{\mathrm{T}} \equiv \sum_{k=0}^{N} R_k. \qquad (2\text{-}39)$$

2.5.4 Applications of the GRE

When a loan is a $1P$-$1R$ loan, the GRE reduces to

$$\frac{R_N}{(1+i)^N} = P_0. \qquad (2\text{-}40)$$

Except for notation, this result is identical to the FRE derived using mathematical induction. As you can see, using the GRE to obtain Equation (2-40) is much simpler than using that approach.

EXAMPLE 2.7 Calculating the Last Payment of a Loan with Two Principles and Three Payments (*E*-Class)

Problem: On New Year's Day the first principle P_0 of a loan is disbursed. Its value is \$1,000, and the interest rate is 8% per year. At the end of the year a payment R_1 of \$500 is received. A year later, a second principle P_2 of \$2,000 is disbursed. One year after that, a payment R_3 is received. The loan is paid off a year later with another payment R_4. *If* $R_4 = 2R_3$, *what is the size of the last payment?*

Solution: Figure 2.8 is the CFD of the loan. The interest rate and the number of interest periods are specified: $i = 0.08$, and $N = 4$. Since $R_4 = 2R_3$, the only unknown in the GRE is R_3:

$$\frac{R_1}{1+i} + \frac{R_3}{(1+i)^3} + \frac{2R_3}{(1+i)^4} = P_0 + \frac{P_2}{(1+i)^2}. \qquad (\text{E}2.7\text{-}1)$$

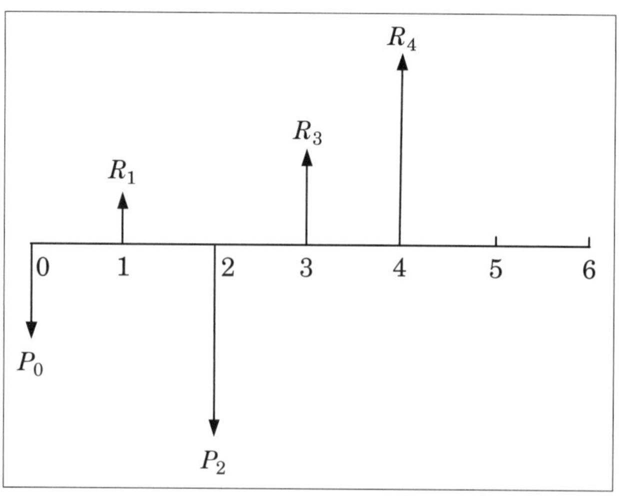

Figure 2.8 **Cash Flows of the Loan** NOTE: Only R_3 is unknown.

Collecting the R_3 terms on one side of Equation (E2.7-1) yields

$$\left(\frac{1}{(1+i)^3}+\frac{2}{(1+i)^4}\right)R_3 = P_0 + \frac{P_2}{(1+i)^2} - \frac{R_1}{1+i}. \tag{E2.7-2}$$

We simplify Equation (2.7-2) as follows: (1) Multiply the first term of the multiplier of R_3 by $(1+i)(1+i)^{-1}$; (2) simplify the multiplier; and (3) solve the resulting equation for R_3:

$$R_3 = \left(\frac{(1+i)^4}{3+i}\right)\left(P_0 + \frac{P_2}{(1+i)^2} - \frac{R_1}{1+i}\right). \tag{E2.7-3}$$

We easily calculate the value of R_3 by substituting the appropriate factor values into Equation (E2.7-3):

$$R_3 = \left(\frac{\underbrace{(F/P, 8\%, 4)}_{1.36049}}{3+0.08}\right)\left(\$1,000 + \$2,000\underbrace{(P/F, 8\%, 2)}_{0.85734} - \$500\underbrace{(P/F, 8\%, 1)}_{0.92593}\right) = \$994.62.$$

The last withdrawl is twice this value:

$$R_4 = 2R_3 = 2 \times \$994.62 = \underline{\underline{\$1,989.24}}.$$

EXAMPLE 2.8 Numerical Confirmation of Equivalence Property 1 (*E*-Class)

Problem: Figure 2.9 displays the cash flows of the loan in Example 2.7. *Confirm that the worths of the principle and payment series at* TP_2 *and* TP_6 *are equal.*

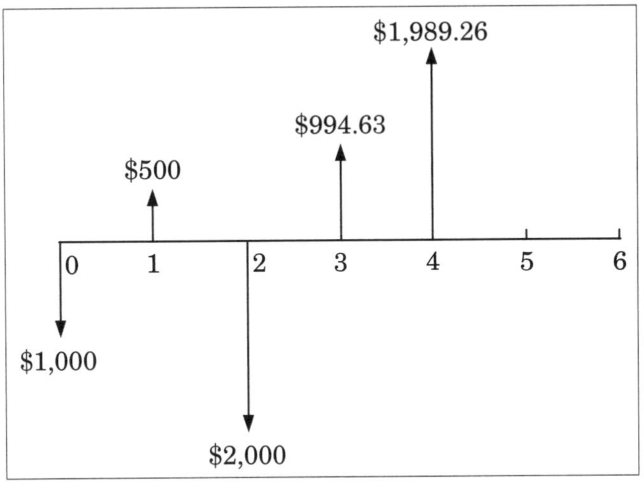

Figure 2.9 ***Magnitudes* of the Annual Cash Flows**

Solution: The worths of the principle and payment series at TP_2 and TP_6 are calculated with the loan interest rate of 8% per year.

Timeline Point 2

This calculates the worths of the principle W_P and payment W_R series at TP_2:

$$W_{P,2} = \$1{,}000\underbrace{(F/P, 8\%, 2)}_{1.16640} + \$2{,}000 = \underline{\underline{\$3{,}166.40}}$$

$$W_{R,2} = \$500\underbrace{(F/P, 8\%, 1)}_{1.08000} + \$994.62\underbrace{(P/F, 8\%, 1)}_{0.92593} + \$1{,}989.24\underbrace{(P/F, 8\%, 2)}_{0.85734} = \underline{\underline{\$3{,}166.40}}$$

NOTES: (1) As expected, the worths of the principle and payment series at TP_2 are *equal*. (2) At TP_2 the worth of R_1 is a future worth, but the worths of R_3 and R_4 are present worths.

Timeline Point 6

NOTES: (1) As expected, the worths of the principle and payment series at TP_6 are also equal. (2) At TP_6 all of the worths are future worths because the last cash flow of the loan ends at TP_4.

$$W_{P,6} = \$1{,}000\underbrace{(F/P, 8\%, 6)}_{1.58687} + \$2{,}000\underbrace{(F/P, 8\%, 4)}_{1.36049} = \underline{\underline{\$4{,}307.85}}$$

$$W_{R,6} = \$500\underbrace{(F/P, 8\%, 5)}_{1.46933} + \$994.62\underbrace{(F/P, 8\%, 3)}_{1.25971} + \$1{,}989.24\underbrace{(F/P, 8\%, 2)}_{1.16640} = \underline{\underline{\$4{,}307.85}}.$$

The three **basic series**—uniform, gradient, and geometric gradient—are often used singly or in some combination to represent a series. The next three sections derive their present and future worth factors. In addition, Section 2.6 derives the uniform worth factors for converting a present cash flow and a future cash flow into *equivalent* uniform series. Sections 2.7 and 2.8 also derive the uniform worth factors for converting a gradient series and a geometric gradient series into an *equivalent* uniform series, respectively.

2.6 UNIFORM SERIES FACTORS & UNIFORM WORTH FACTORS of a CASH FLOW

2.6.1 Present Worth Factor

Present Worth of the Series

The cash flows of a **uniform series** have the same magnitude A, which is called the **metric**[4] of the series. Figure 2.10 is the CFD of a uniform series with N cash flows. (Most consumer loans except credit card debt have a single principle at TP_0 followed by a uniform series of payments.) Since the cash flows of a uniform series have the same magnitude A, the **present worth of a uniform series** PW_A is given by the following sum:

$$PW_A = \frac{A}{1+i} + \frac{A}{(1+i)^2} + \cdots + \frac{A}{(1+i)^N}.$$

---

[4]The **metric of *any* basic series** is a parameter that defines the magnitudes of its cash flows. The metric of a uniform series is A, the constant magnitude of its cash flows.

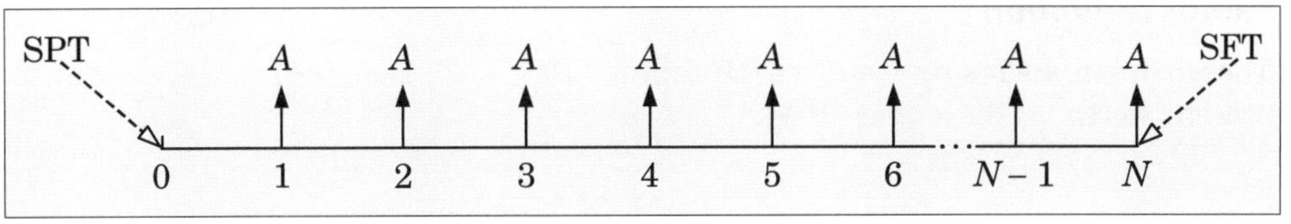

Figure 2.10 **CFD of a Uniform Series**

$$PW_A = \sum_{k=1}^{N} \frac{A}{(1+i)^k}. \tag{2-41}$$

The **series present time** (SPT) is *defined* as *one* interest period *before* the timeline point of the first cash flow of the series, and the **series future time** (SFT) is *defined* as the timeline point of its *N*th cash flow.

Present Worth Equation

To evaluate the sum in Equation (2-41), we note that the **sequence** of the ***N* present worths** *a*

$$a \equiv \frac{A}{1+i}, \frac{A}{(1+i)^2}, \ldots, \frac{A}{(1+i)^N} \tag{2-42}$$

is a **geometric sequence** because the *ratios* of its *consecutive* terms have the *same* value *r*:

$$r \equiv \frac{A}{(1+i)^{k+1}} \left[\frac{A}{(1+i)^k} \right]^{-1} = \frac{1}{1+i} \qquad k = 1, 2, \ldots, N-1. \tag{2-43}$$

The **sum S_N of the terms of *any* geometric sequence of *N* terms** is given by[5]

$$S_N = a_1 \frac{1-r^N}{1-r} \qquad r \neq 1, \tag{2-44}$$

where a_1 is the first term of the geometric sequence.

Substituting the values of a_1 from Equation (2-42) and *r* from Equation (2-43) into Equation (2-44) yields the present worth of the uniform series:

$$PW_A = \sum_{k=1}^{N} \frac{A}{(1+i)^k} = \frac{A}{1+i} \frac{1-(1+i)^{-N}}{1-(1+i)^{-1}}$$

$$\boxed{PW_A = A \frac{1-(1+i)^{-N}}{i}.} \tag{2-45}$$

[5] Larson, R. and Hostetler, R.P. *College Algebra*, 5th ed., p. 601. Boston: Houghton Mifflin, 2001.

Factor Definition

The **uniform series present worth factor** $(P/A, i, N)$ is *defined* as the normalized present worth[6] of the series—$PW_A A^{-1}$:

$$\boxed{(P/A, i, N) \equiv \frac{PW_A}{A} = \frac{1-(1+i)^{-N}}{i}.} \tag{2-46}$$

Hence, we calculate PW_A with the value of $(P/A, i, N)$:

$$\boxed{PW_A = A(P/A, i, N).} \tag{2-47}$$

Computation Note: *A present worth comparison of two or more economic alternatives is a* ***PWC-Class*** *problem.*

EXAMPLE 2.9 Choosing Between Two Sweepstakes Prizes (*PWC*-Class)

Problem: At the end of 2006, American Express assigned unique numbers to its cardmembers. If a cardmember returned the number before a specified date, and if it matched a preselected number, the cardmember won \$1 million dollars. The winner had a choice of receiving \$501,000 immediately or 30 annual payments of \$33,333.33—which equals \$999,999.90—starting immediately. *If the prevailing interest rate is assumed to be 4% per year for 29 years, which prize is worth more?*

Solution: *One way to compare the worths of alternatives is to compare their present worths at a common time.* Since the \$501,000 alternative occurs now, we compare it to the present worth of the 30 annual payments. This is the sum of the immediate payment of \$33,333.33 and the present worth of the 29 annual payments of \$33,333.33. Since $i = 0.04$ per year, the present worth of the 30 annual payments is given by

$$PW_{\text{payments}} = \$33{,}333.33\left[1 + \underbrace{(P/A, 4\%, 29)}_{16.98371}\right] = \underline{\underline{\$599{,}456.94}}.$$

The payment series ***worth*** *is \$98,456.94 greater than that of the lump sum.*

However, the following questions should be answered in the prize comparison:

1. Is there an immediate use for the \$501,000?
2. Either prize will be taxed, but the income tax rate on \$501,000 is greater than that on \$33,333.33. How does this affect the comparison?
3. Interest rates vary over time. How will this affect the present worth of the 29 future payments?
4. How will inflation affect the purchasing power of the 29 future payments?

---

[6] The **normalized worth of any basic series** is its worth divided by its metric.

2.6.2 Future Worth Factor

Future Worth Equation

The **future worth of a uniform series** FW_A follows from Equations (2-29) and (2-45):

$$FW_A = PW_A(1+i)^N$$
$$= A\frac{1-(1+i)^{-N}}{i}(1+i)^N.$$

Therefore,

$$\boxed{FW_A = A\frac{(1+i)^N - 1}{i},} \quad (2\text{-}48)$$

where the **series future time** SFT has been previously *defined* as the timeline point of the Nth cash flow TP_N (see Figure 2.10).

Factor Definition

The **uniform series future worth factor** $(F/A, i, N)$ is *defined* as the normalized future worth of the series —$FW_A A^{-1}$:

$$\boxed{(F/A, i, N) \equiv \frac{FW_A}{A} = \frac{(1+i)^N - 1}{i}.} \quad (2\text{-}49)$$

Hence, we calculate FW_A with the value of $(F/A, i, N)$:

$$\boxed{FW_A = A(F/A, i, N).} \quad (2\text{-}50)$$

EXAMPLE 2.10 Calculating a Christmas Club Balance (*E*-Class)

Problem: On January 31st a customer opens bank account, with an APR of 3% compounded monthly, and makes the first of 11 equal end-of-month deposits of \$100. *How much money can be withdrawn from the account on December 1st?*

Solution: The withdrawl is just the future worth of the 11 deposits. To calculate it, we must first determine the value of $i\%$. It follows from the APR that $p = 1$ mo. Hence, $m = 12$. We use Equation (2-8) to calculate $i\%$:

$$i\% = \frac{APR}{m} = \frac{3}{12} = 0.25\% \text{ per month.}$$

Hence, the withdrawl amount is given by

$$FW_A = \$100\left[\underbrace{(F/P, 0.25\%, 10)}_{1.02528} + \underbrace{(F/A, 0.25\%, 10)}_{10.11325}\right] = \underline{\underline{\$1,113.85.}}$$

2.6.3 Uniform Worth of a Present Cash Flow

Definition

When a loan consists of a *specified* principle P followed by a uniform series of N payments, we need to calculate the *metric* A of the uniform series of N cash flows that is equivalent P. Hence, the **uniform worth of a present cash flow** UW_P is defined as the *metric* A of a uniform series that is equivalent P:

$$PW_A = P, \tag{2-51}$$

where PW_A is given by Equation (2-45):

$$PW_A = A\frac{1-(1+i)^{-N}}{i}. \tag{2-45}$$

Equating Equations (2-51) and (2-45) and solving for A and *defining* UW_P as A yields:

$$\boxed{UW_P \equiv A = P\frac{i}{1-(1+i)^{-N}}.} \tag{2-52}$$

Capital Recovery Factor

The **capital recovery factor**[7] $(A/P,i,N)$ is *defined* as the normalized uniform worth of P, which is $UW_P P^{-1}$:

$$\boxed{(A/P,i,N) \equiv \frac{UW_P}{P} = \frac{i}{1-(1+i)^{-N}}.} \tag{2-53}$$

Hence, we calculate UW_P with the value of $(A/P,i,N)$:

$$\boxed{UW_P \equiv A = P(A/P,i,N).} \tag{2-54}$$

It follows from Equations (2-46) and (2-53) that *the capital recovery factor is the inverse of the uniform series present worth factor:*

$$(A/P,i,N) = (P/A,i,N)^{-1}. \tag{2-55}$$

1P-A Loan

In this book a ***1P-A* loan** is *defined* as a loan with one principle and a uniform series of payments. A CFD of 1*P-A* loan is displayed in Figure 2.11.

[7] **Capital recovery** is the recovery of a present amount of invested money (capital) with interest as a uniform series of receipts. The recovery of the principle of an automobile loan with a uniform series of payments is an example of capital recovery.

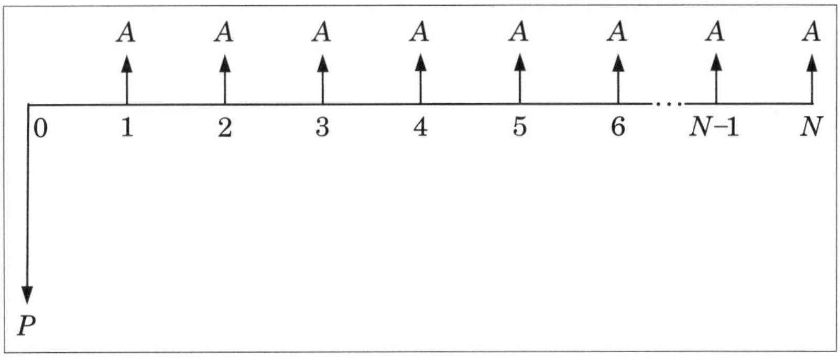

Figure 2.11 **CFD of a 1*P*-*A* Loan**

Computation Note: *Since the payment series of a 1P-A loan is equivalent to the principle, we use Equation (2-54) to calculate A with the value of* $(A/P, i, N)$.

EXAMPLE 2.11 Monthly Payment for a Used Corvette (*E*-Class)

Problem: A salesman found a mint used Corvette that costs \$27,000. If he puts down \$3,000, he can get a four-year bank loan for the balance. *If the APR is 6% compounded monthly, what is the monthly payment of the loan?*

Solution: This is a 1*P*-*A* loan whose term and principle are specified: $T = 4\,\text{a}$, and $P = \$27{,}000 - \$3{,}000 = \$24{,}000$. To calculate the monthly payment *A*, we must first determine the values of $i\%$ and *N*.

It follows from the *APR* that $p = 1\,\text{mo}$. Hence, $m = 12$. We use Equation (2-8) to calculate $i\%$:

$$i\% = \frac{APR}{m} = \frac{6}{12} = 0.5\% \text{ per month.}$$

Equation (2-1) is used to calculate N:

$$N = \frac{T}{p} = \frac{4\,\text{a} \times 12\,\text{mo}\cdot\text{a}^{-1}}{1\,\text{mo}} = 48.$$

The uniform series of payments is equivalent to the principle. Hence,

$$UW_P = A = \$24{,}000\underbrace{(A/P, 0.5\%, 48)}_{0.023485} = \underline{\underline{\$563.64.}}$$

Total and Normalized Total Interests of a 1P-A Loan

The total interest of any loan is given by Equation (2-4):

$$I_T = R_T - P_T, \tag{2-4}$$

Hence, the **total interest** I_T of a 1*P*-*A* loan is given by

$$I_T = NA - P. \tag{2-56}$$

Dividing Equation (2-56) by *P* yields the **normalized total interest** $I_T P^{-1}$ of a 1*P*-*A* loan:

$$\frac{I_T}{P} = N\frac{A}{P} - 1. \tag{2-57}$$

EXAMPLE 2.12 Comparing the Total Interests of a 15-Year Mortgage and a 30-Year Mortgage (*E*-Class)

Problem: A bank offers 15-year mortgages and 30-year mortgages, and each has an *APR* of 6.5% compounded monthly. *(1) If the principle of a mortgage is \$100,000, what is the monthly payment for each term. (2) What are the total interests (rounded to the dollar) earned by the loans?*

Solution: The mortgages are 1*P*-*A* loans whose common principles and terms are specified: $P = \$100{,}000$, $T_{15} = 15\,\text{a}$ and $T_{30} = 30\,\text{a}$. To calculate the corresponding monthly payments A_{15} and A_{30}, we must first determine the value of $i\%$ and the values of N_{15} and N_{30}. We use Equation (2-8) to calculate $i\%$:

$$i\% = \frac{APR}{m} = \frac{6.5}{12} = 0.54167\% \text{ per month.}$$

Equation (2-1) is used to calculate the number of interest periods in the terms:

$$N_{15} = \frac{T_{15}}{p} = \frac{15\,\text{a} \times 12\,\text{mo·a}^{-1}}{1\,\text{mo}} = 180$$

$$N_{30} = \frac{T_{30}}{p} = \frac{30\,\text{a} \times 12\,\text{mo·a}^{-1}}{1\,\text{mo}} = 360.$$

We use Equation (2-54) to calculate the monthly payments:

$$A_{15} = \$100,000\underbrace{(A/P, 0.54167\%, 180)}_{0.0087111} = \underline{\underline{\$871.11}}.$$

$$A_{15} = \$100,000\underbrace{(A/P, 0.54167\%, 360)}_{0.0063207} = \underline{\underline{\$632.07}}.$$

Equation (2-54) is used to calculate the total interest of each mortgage:

$$I_{T_{15}} = N_{15}A_{15} - P = 180 \times \$871.11 - \$100{,}000 = \underline{\underline{\$56{,}799.80}}$$

$$I_{T_{30}} = N_{30}A_{30} - P = 360 \times \$632.07 - \$100{,}000 = \underline{\underline{\$127{,}545.20}}.$$

The total interest of the 30-year mortgage is more than twice that of the 15-year mortgage! Remember this when you buy a house.

Limiting Forms of the Capital Recovery Factor

The capital recovery factor has two limiting forms:

$$\lim_{i\to 0}\left(A/P,i,N\right)=\frac{1}{N} \tag{2-58}$$

and

$$\lim_{N\to\infty}\left(A/P,i,N\right)=i. \tag{2-59}$$

Equation (2-58) is an obvious result. If the interest rate is zero, we *expect* that $A = PN^{-1}$. Calculating the limit just confirms our intuition. To do so, we note that

$$\lim_{i\to 0}\left(A/P,i,N\right)=\lim_{i\to 0}\frac{i}{1-(1+i)^{-N}}=\frac{0}{0},$$

which is indeterminate. Hence, we use L'Hôpital's Rule to obtain the limit:

$$\lim_{i\to 0}\left(A/P,i,N\right)=\lim_{i\to 0}\left[\frac{\frac{d}{di}i}{\frac{d}{di}\left(1-(1+i)^{-N}\right)}\right]=\lim_{i\to 0}\left[\frac{1}{N(1+i)^{-(N-1)}}\right]=\frac{1}{N}. \tag{2-58}$$

Equation (2-59) is obtained by noting that the term $(1+i)^{-N}$ in the denominator of Equation (2-53) approaches zero as N approaches infinity. Hence,

$$UW_{P,\infty}\equiv\lim_{N\to\infty}UW_P=A_\infty=Pi. \tag{2-60}$$

Each payment is just the interest earned by P. Hence, ***the balance is always P, and the loan is never paid off!*** This type of loan is called a **perpetuity**. A lender may convert a bad loan into a perpetuity to avoid a complete default by the borrower. This often happens when the lender and borrower are countries.

Usage Note*: It does not take an infinite number of interest periods for a loan to behave like a perpetuity. For example, if its interest rate is 0.15, the capital recovery factor is 0.15003 when N = 60. For smaller interest rates N is larger and vice versa.*

2.6.4 Uniform Worth of a Future Cash Flow

Definition

The **uniform worth of a future cash flow** UW_F at TP_N is *defined* as the metric of a uniform series of N cash flows that is *equivalent* to F. The uniform series and F are equivalent when their future worths are equal. That is,

$$FW_A = F, \tag{2-61}$$

where FW_A is given by Equation (2-48):

$$FW_A = A\frac{(1+i)^N - 1}{i}. \tag{2-48}$$

Equating Equations (2-61) and (2-48) and solving for A and *defining* UW_F as A yields

$$\boxed{UW_F \equiv A = F\frac{i}{(1+i)^N - 1}.} \tag{2-62}$$

Sinking Fund Factor

The **sinking fund factor**[8] $(A/F, i, N)$ is *defined* as the normalized uniform worth of F, $UW_F F^{-1}$:

$$\boxed{(A/F, i, N) \equiv \frac{UW_F}{F} = \frac{i}{(1+i)^N - 1}.} \tag{2-63}$$

Hence, we calculate UW_F with the value of $(A/F, i, N)$:

$$\boxed{UW_F \equiv A = F(A/F, i, N).} \tag{2-64}$$

It follows from Equations (2-49) and (2-63) that *the sinking fund factor is the inverse of the uniform series future worth factor:*

$$(A/F, i, N) = (F/A, i, N)^{-1}. \tag{2-65}$$

EXAMPLE 2.13 Creating a Retirement Plan That Is a Perpetuity (*WC*-Class)

Problem: An engineer wants to retire when she is 59. She is planning her retirement income, so that on her 60th birthday she starts receiving $75,000 annually for the rest of her life from a conservative bond fund that has historically earned 5% per year. To provide this annual income, she plans to invest a constant amount of money annually in a mutual fund starting on her 25th birthday and ending on her 59th. *What is the size of her annual investment if the mutual fund has an expected average rate of return of 8% per year?*

Solution: Figure 2.12 shows the cash flows of the problem. The future worth F of the uniform series of investment cash flows must be the principle P invested in the bond fund. That is, *the worths of the investment and retirement series at* TP_{59} *must be equal.*

---

[8] A **sinking fund** is an interest-bearing account into which an organization periodically deposits money to pay off a debt at a future time.

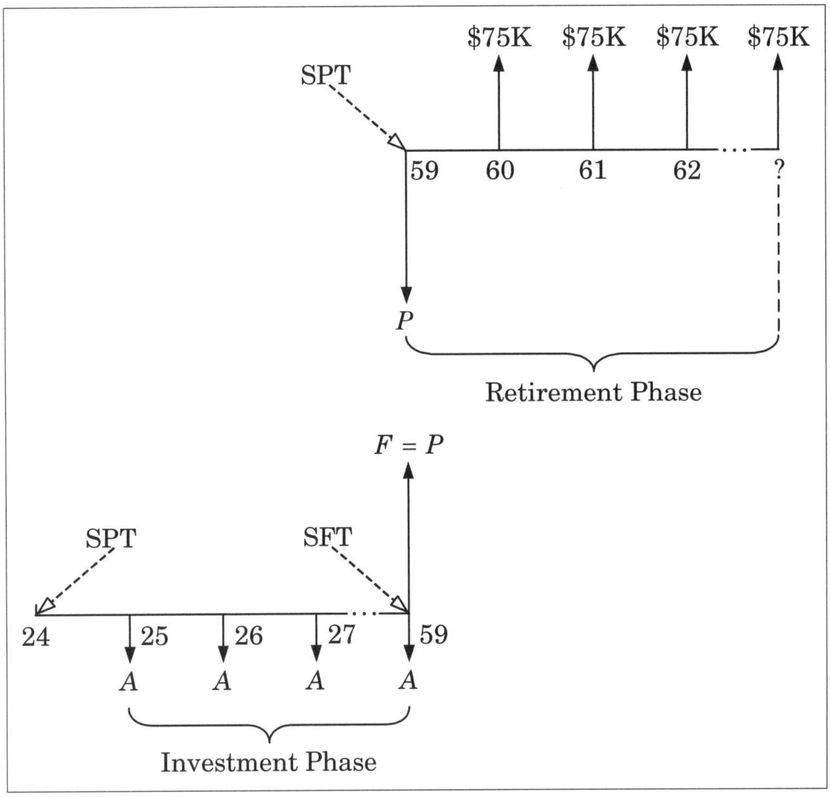

Figure 2.12 **CFD of the Engineer's Retirement Plan**

However, *these series are not equivalent series because their interest rates are different.* Hence, ***their worths at any other timeline point will not be equal!***

The investment P in the bond fund is assumed to be a perpetuity whose annual interests are each \$75,000. Since the interest rate earned by the bond fund is 5% per year, it follows from Equation (2-60) that

$$P = \frac{A_\infty}{i} = \frac{\$75,000}{0.05} = \$1,500,000.$$

The future worth F of the mutual fund investment, which is a sinking fund, must equal P. Substituting the values of F and N, which is given by

$$N = \text{SFT} - \text{SPT} = 59 - 24 = 35,$$

into Equation (2-64) yields the size of the annual investment:

$$UW_F = A = \$1,500,000\underbrace{(A/F, 8\%, 35)}_{0.0058033} = \underline{\underline{\$8,704.95.}}$$

This annual investment will provide an annual retirement income that is over 8.6 times as large. In addition, the engineer's heirs will receive the principle of \$1,500,000.

2.6.5 Shifted Basic Series

Present Worth

As stated in Subsection 2.6.1, there are three basic series. Here "B" represents the metric (A, G, or A_1) of one of them. A **shifted basic series** is one whose SPT does not coincide with the PT of a problem. Figure 2.13 shows a uniform series that is shifted one interest period into the *future*. NOTE: The PT is TP_{23}, but the SPT is TP_{24}.

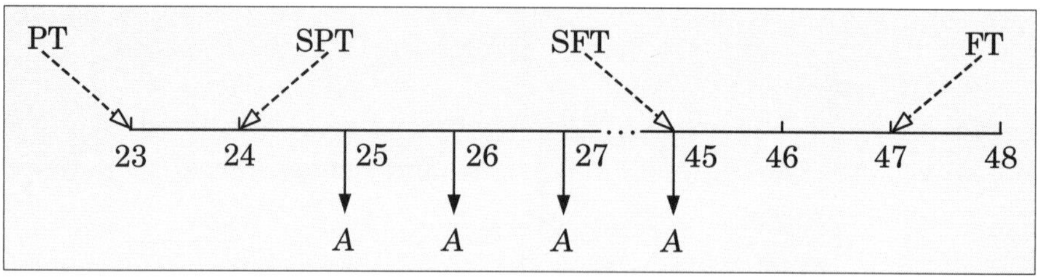

Figure 2.13 **Uniform Series Shifted One Interest Period Into the Future**

Usage Note: *The equations for the present and future worths of a basic series refer to their worths at the SPT and SFT, respectively.*

To calculate the **present worth** of a **shifted basic series** $PW_{B,\mathrm{PT}}$ at an *arbitrary* PT, we use the following form of Equation (2-28):

$$\boxed{PW_{B,\mathrm{PT}} = PW_B(1+i)^{-\delta_{\mathrm{PT}}},} \tag{2-66}$$

where PW_B is the present worth of the basic series at its SPT and δ_{PT} is the difference of the series present time and the actual present time:

$$\delta_{\mathrm{PT}} \equiv TP_{\mathrm{SPT}} - TP_{\mathrm{PT}}. \tag{2-67}$$

When $\delta_{\mathrm{PT}} > 0$, as in Figure 2.13,

$$\boxed{PW_{B,\mathrm{PT}} = PW_B\left(P/F, i, \delta_{\mathrm{PT}}\right).} \tag{2-68}$$

When $\delta_{\mathrm{PT}} < 0$,

$$\boxed{PW_{B,\mathrm{PT}} = PW_B\left(F/P, i, -\delta_{\mathrm{PT}}\right).} \tag{2-69}$$

EXAMPLE 2.14 Equivalent Shifted Uniform Series of a Loan (*E*-Class)

Problem: A furniture store offers two payment plans for a pair of couches:

- Plan X: three equal annual payments of \$750 that start now
- Plan Y: four equal annual payments that start in three years

(1) If the financing interest rate is 12% per year, what is the annual payment of Plan Y that makes the two payment plans equivalent? (2) What is the difference of the total payments of each plan?

Solution: Figure 2.14 shows the CFDs of the two payment plans. Both payment series are *shifted* uniform series. They are equivalent if their present worths at the PT, which is TP_0, are equal.

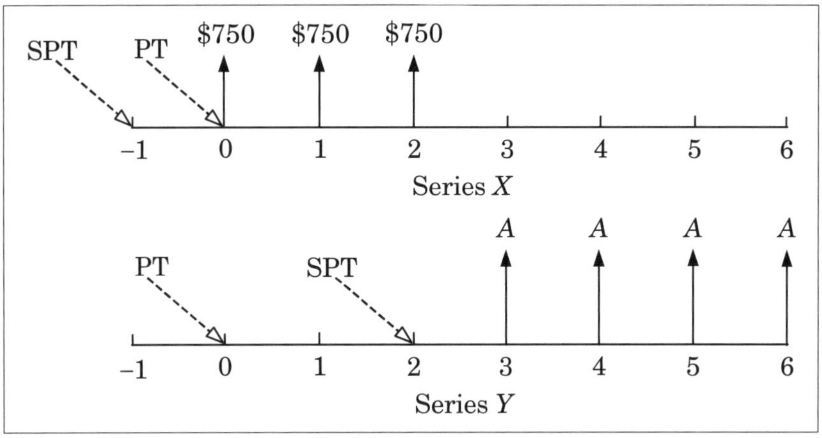

Figure 2.14 **CFDs of the Two Payment Plans of the Furniture Loan**

To determine the value of A, we

1. Calculate the present worth $PW_{X,PT}$ of series X at the PT.
2. Calculate the present worth $PW_{Y,PT}$ of series Y at the PT. It will contain the unknown metric A.
3. Equate $PW_{Y,PT}$ to $PW_{X,PT}$ and solve for A.

Step 1

Since $\delta_{PT} = TP_{SPT} - TP_{PT} = -1 - 0 = -1 < 0$, we use Equation (2-69) to calculate $PW_{X,PT}$:

$$PW_{X,PT} = \$750\underbrace{(P/A,12\%,3)}_{2.40183}\underbrace{(F/P,12\%,1)}_{1.12000} = \$2{,}017.54.$$

Step 2

Since $\delta_{PT} = TP_{SPT} - TP_{PT} = 2 - 0 = 2 > 0$, we use Equation (2-68) to calculate $PW_{Y,PT}$:

$$PW_{Y,PT} = A\underbrace{(P/A,12\%,4)}_{3.03735}\underbrace{(P/F,12\%,2)}_{0.79719} = 2.42135A$$

Step 3

Equate $PW_{Y,PT}$ to $PW_{X,PT}$ and solve for A:

$$A = \frac{\$2{,}017.54}{2.42135} = \underline{\underline{\$833.23.}}$$

The difference of the total payments of series Y and series X is large:

$$4 \times \$833.23 - 3 \times \$750 = \underline{\underline{\$1,082.92}}.$$

Future Worth

A basic series whose SFT does not coincide with the FT of a problem is also a **shifted basic series**. Figure 2.15 shows a shifted uniform series whose SFT is TP_{45} and FT is TP_{49}.

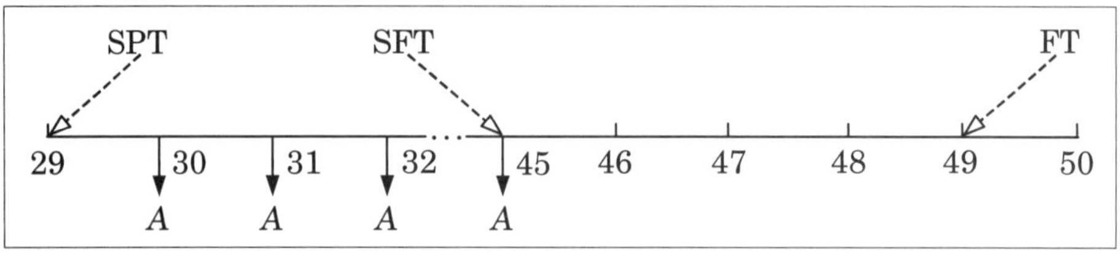

Figure 2.15 **Uniform Series Whose SFT Does Not Coincide with the FT**

Usage Note: The equations for the present and future worths of a basic series refer to their worths at the SPT and SFT, respectively.

To calculate the **future worth** of a **shifted basic series** $FW_{B,\mathrm{FT}}$ at an *arbitrary* FT, we use an equation that is analogous to Equation (2-66):

$$\boxed{FW_{B,\mathrm{FT}} = FW_B(1+i)^{\delta_{\mathrm{FT}}},} \tag{2-70}$$

where FW_B is the future worth of the basic series at its SFT, and

$$\delta_{\mathrm{FT}} \equiv TP_{\mathrm{FT}} - TP_{\mathrm{SFT}}. \tag{2-71}$$

When $\delta_{\mathrm{FT}} > 0$, as in Figure 2.15,

$$\boxed{FW_{B,\mathrm{FT}} = FW_B\left(F/P, i, \delta_{\mathrm{FT}}\right).} \tag{2-72}$$

When $\delta_{\mathrm{FT}} < 0$,

$$\boxed{FW_{B,\mathrm{FT}} = FW_B\left(P/F, i, -\delta_{\mathrm{FT}}\right).} \tag{2-73}$$

EXAMPLE 2.15 Investing to Provide a College Education (*E*-Class)

Problem: To provide $100,000 for the college education of his baby daughter, a fireman wants to invest a constant amount of money annually in a mutual fund starting on his 30th birthday and ending on his 45th. He will leave the money in the mutual fund until his 50th birthday. At that time he will make the first of four $25,000 annual withdrawals for his daughter's college education. *If the expected average rate of return of the mutual fund is 7% per year, what is the size of the annual investment that will provide a final withdrawl that reduces the fund balance to zero?*

Solution: Figure 2.16 is the CFD of the investment, which is essentially a loan to the mutual fund.

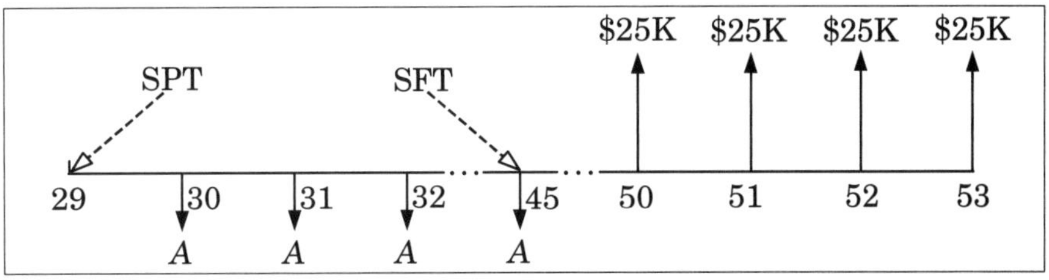

Figure 2.16 **CFD for Example 2.15**

Since *the principle and payment series of a loan are equivalent*, we can equate their worths at *any* point on the timeline, such as TP_{49}, to calculate A:

$$\underbrace{A\underbrace{(F/A,7\%,16)}_{27.88805}\underbrace{(F/P,7\%,4)}_{1.31080}}_{\text{Future Worth of Principles at } TP_{49}} = \underbrace{\$25{,}000\underbrace{(P/A,7\%,4)}_{3.38721}}_{\text{Present Worth of Payments at } TP_{49}}$$

Hence,

$$A = \frac{\$25{,}000 \times 3.38721}{27.88805 \times 1.31080} = \underline{\underline{\$2,316.47.}}$$

2.7 GRADIENT SERIES FACTORS

2.7.1 Present Worth Factor

Present Worth of the Series

The cash flows of a **gradient series** *increase* from period-to-period by a *constant* amount G, which is the **metric** of the series. Figure 2.17 is a CFD of a gradient series with $N-1$ cash flows. Unlike a uniform series, *the first cash flow of a gradient series occurs at the end of the second interest period.*

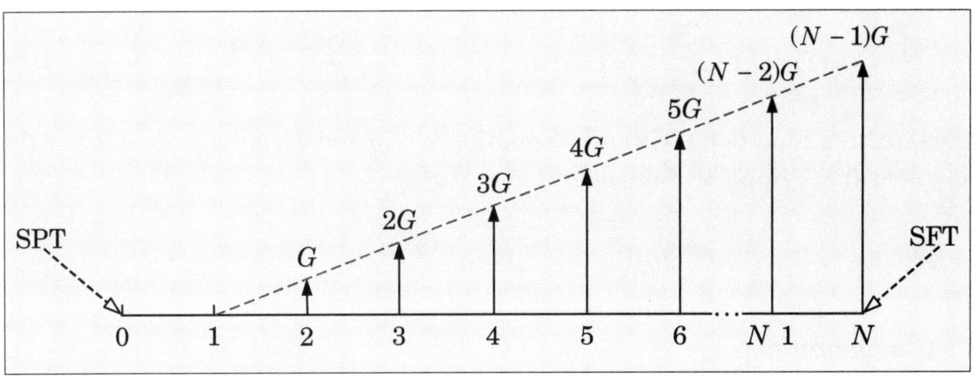

Figure 2.17 **CFD of a Gradient Series**

The **present worth of a gradient series** PW_G is given by the following sum:

$$PW_G = \frac{G}{(1+i)^2} + \frac{2G}{(1+i)^3} + \cdots + \frac{(N-1)G}{(1+i)^N} = \sum_{k=2}^{N} \frac{(k-1)G}{(1+i)^k}, \tag{2-74}$$

where the **series present time** SPT is *defined* as *two* interest periods *before* the timeline point of the first cash flow, and the **series future time** SFT is *defined* as the timeline point of the $(N-1)$th cash flow.

Present Worth Equation

The sequence of the present worths in Equation (2-74) is *not* a geometric sequence because the ratios of its consecutive terms have different values. However, we can derive a geometric sequence from Equation (2-74). First we multiply it by $(1+i)^N$:

$$PW_G(1+i)^N = (N-1)G + (N-2)G(1+i) + (N-3)G(1+i)^2 + \cdots + G(1+i)^{N-2} \tag{2-75}$$

Next, we multiply Equation (2-75) by $1+i$:

$$PW_G(1+i)(1+i)^N = (N-1)G(1+i) + (N-2)G(1+i)^2 + \cdots + G(1+i)^{N-1} \tag{2-76}$$

The multipliers of the $(1+i)^k$ terms in Equations (2-75) and (2-76) differ by G. Hence, subtracting Equation (2-75) from Equation (2-76) yields

$$PW_G i(1+i)^N = \left[G + G(1+i) + G(1+i)^2 + \cdots + G(1+i)^{N-1}\right] - GN. \tag{2-77}$$

The **sequence** of the terms in the brackets is a **geometric series** with the parameter values

$$\begin{aligned} a_1 &= G \\ r &= G(1+i)^{k+1}\left[G(1+i)\right]^{-k} = (1+i). \end{aligned} \tag{2-78}$$

Hence, the sum of its terms is given by Equation (2-44):

$$G + G(1+i) + G(1+i)^2 + \cdots + G(1+i)^{N-1} = G\frac{1-(1+i)^N}{1-(1+i)} = G\frac{(1+i)^N - 1}{i}. \tag{2-79}$$

Substituting Equation (2-79) into Equation (2-77) and dividing both sides of the equation by $i(1+i)^N$ yields the equation for PW_G:

$$\boxed{PW_G = G\frac{(1+i)^N - Ni - 1}{i^2(1+i)^N}.} \tag{2-80}$$

Factor Definition

The **gradient series present worth factor** $(P/G, i, N)$ is *defined* as the normalized present worth of the series—$PW_G G^{-1}$:

$$\boxed{\left(P/G,i,N\right) \equiv \frac{PW_G}{G} = \frac{(1+i)^N - Ni - 1}{i^2(1+i)^N}.} \tag{2-81}$$

Hence, we calculate PW_G with the value of $\left(P/G,i,N\right)$:

$$\boxed{PW_G = G\left(P/G,i,N\right).} \tag{2-82}$$

Using PW_G in a Least-Cost Study

A **least-cost study** is a ***WC*-Class** problem that identifies the least-cost alternative among several **cost alternatives**—alternatives with cash flows that are *predominately* disbursements. To compare cost alternatives, we apply one of the **Worth Comparison Rules**:

Worth Comparison Rules
W-Rule: To compare the worths of two or more alternatives their worths must be calculated at a common time (present or future).
OR
AW-Rule: To compare the worths of two or more alternatives, their uniform worths must be calculated on a common annual timespan.

Computation Notes: *(1) One way to perform a least-cost study is to apply the W-Rule at the present time to find the cost alternative with the smallest positive present worth of its cash flows (a* ***PWC-Class*** *problem). NOTE: Disbursements are defined as positive cash flows to avoid cluttering the calculations with negative numbers. (2) The interest rate used to calculate the present worths of the cost alternatives is called the* ***minimum attractive rate of return*** *(MARR) r^*.*[9]

EXAMPLE 2.16 Finding the Least-Cost Compressor: A Present Worth Comparison When $ST = L = 10$ a (*PWC*-Class)

Problem: A chemical company requires a new compressor for a 10-year **service time** *ST*. Table 2.2 lists the economic data for compressors *X* and *Y*. NOTE: A gradient series is often used to model the annual **maintenance costs** of equipment.[10] *Which compressor is the least-cost alternative when the MARR% is 20% per year?*

TABLE 2.2 Economic Data for the Compressors

DATA TYPE	*X*	*Y*
Purchase Price: *P*	\$80,000	\$85,000
Salvage Value[1]: *S*	\$4,000	\$4,500
Annual Energy Cost: *A*	\$3,000	\$2,200

---

[9] The MARR is the minimum interest rate that a *proposed* investment must earn before a company will consider funding it. Subsection 4.1.1 provides additional information.

[10] The maintenance cost for the first year is normally covered by a manufacturer's warranty.

TABLE 2.2 Economic Data for the Compressors *(continued)*

Annual Maintenance Cost Increment: G	$750	$460
Service Time: ST	10 a	

[1]The **salvage value** of a physical object is its value at the end of its **life** L: the timespan on which it meets its performance specifications.

Solution: Table 2.3 shows the calculations for $N = 10$ annual interest periods. We use a **present worth table** to present them because it clearly displays and contrasts the present worths of each compressor's **cash flow elements**—its series and single cash flows. The ***PRESENT WORTH COMPARISON Calculator***, an Excel spreadsheet created by the author and located in the *Book Templates* file, *automates* present worth comparisons of cost alternatives. *It is described in Subsection 5.1.2.*

TABLE 2.3 Comparing the Costs of Compressors X and Y: $ST = L =$ 10 a; $r^{*}\% = 20\%$ per Year[1]

Worth Type	*Present Worth Table for Compressors X and Y*	
	X	***Y***
PW_P	$80,000	$85,000
PW_S	$(\$4,000)\underbrace{(P/F, 20\%, 10)}_{0.16151} = (\$646)$	$(\$4,500)\underbrace{(P/F, 20\%, 10)}_{0.16151} = (\$727)$
PW_A	$\$3,000\underbrace{(P/A, 20\%, 10)}_{4.19247} = \$12,577$	$\$2,200\underbrace{(P/A, 20\%, 10)}_{4.19247} = \$9,223$
PW_G	$\$750\underbrace{(P/G, 20\%, 10)}_{12.88708} = \$9,665$	$\$460\underbrace{(P/G, 20\%, 10)}_{12.88708} = \$5,928$
ΣPW	**$101,596**	**$99,424**

[1] *Worth entries in cost comparison tables or spreadsheets are always rounded to the dollar because of the uncertainty in predicting their values over many interest periods.*

Computation Note: *Negative numbers are denoted by enclosing their magnitudes in parentheses. This is one of the four ways that Excel represents a negative number.*

Over 10 years compressor X is $2,172 more expensive than compressor Y. Even though the purchase price of compressor Y is greater than that of compressor X, it is the least-cost alternative because of its lower energy and maintenance costs and higher salvage value.

The decision to purchase one of the compressors also depends on its

- Performance Specifications
- Delivery Time
- Reliability
- Vendor Reliability
- Parts Availability

These items are called **irreducible considerations** because it is difficult to assign costs to them. But a preponderance of favorable irreducible considerations for a more costly alternative may override its economic disadvantage.

Chapters 4 and 5 show how to use uniform worth and present worth least-cost studies, respectively, to compare two or more cost alternatives.

2.7.2 Future Worth Factor

Future Worth Equation

The **future worth of a gradient series** FW_G follows from Equations (2-29) and (2-80):

$$
\begin{aligned}
FW_G &= PW_G(1+i)^N \\
&= G\,\frac{(1+i)^N - Ni - 1}{i^2(1+i)^N}(1+i)^N,
\end{aligned}
$$

which simplifies to

$$
\boxed{FW_G = G\left(\frac{1}{i}\,\frac{(1+i)^N - 1}{i} - \frac{N}{i}\right).} \tag{2-83}
$$

We obtain a simple interesting result by noting that the second quotient in this equation is jūst $FW_A A^{\,1}$(see Equation (2-48)). Hence,

$$
\boxed{FW_G = G\left(\frac{1}{i}\,\frac{FW_A}{A} - \frac{N}{i}\right),} \tag{2-84}
$$

where the **series future time** SFT has been previously *defined* as the timeline point of the $(N-1)$th cash flow of the series (see Figure 2.17).

Factor Definition

The **gradient series future worth factor** $(F/G,i,N)$ is *defined* as the normalized future worth of the series $FW_G G^{-1}$:

$$
\boxed{(F/G,i,N) \equiv \frac{FW_G}{G} = \frac{(F/A,i,N) - N}{i}.} \tag{2-85}
$$

Hence, we calculate FW_G with the value[11] of $(F/G,i,N)$:

$$
\boxed{FW_G = G(F/G,i,N).} \tag{2-86}
$$

2.7.3 Uniform Worth of a Gradient Series

Definition

In Chapter 4 we will need to transform a gradient series into an *equivalent* uniform series with the *same* SPT and number of interest periods. Figure 2.18 shows this transformation.

The **uniform worth of the gradient series** UW_G is defined as the *metric* A of the equivalent uniform series. The two series are equivalent when their present worths are equal:

[11] Most textbooks on engineering economics do not tabulate this factor, but it is convenient to do so. Hence, the ***12-FACTOR Calculator*** provides values of $(F/G,i,N)$.

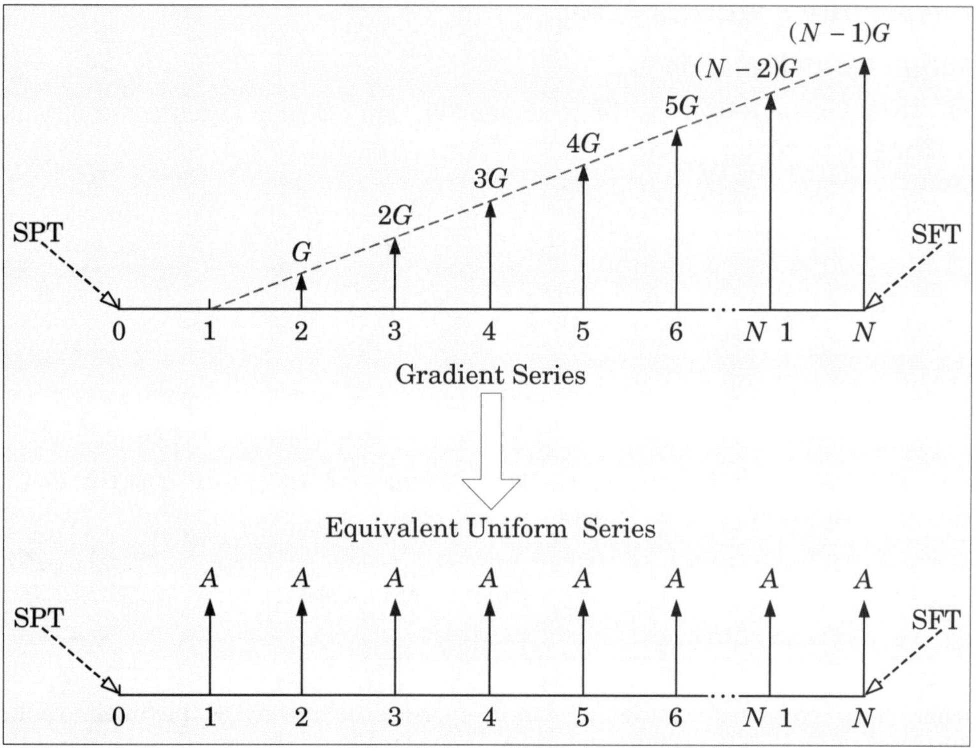

Figure 2.18 **Transforming a Gradient Series to an *Equivalent* Uniform Series**

$$PW_A = PW_{G.}$$

Substituting the expressions for PW_A and PW_G from Equations (2-45) and (2-80), respectively, into the equality yields

$$A\frac{1-(1+i)^{-N}}{i} = G\frac{(1+i)^N - Ni - 1}{i^2(1+i)^N}.$$

Hence, the value of A is given by

$$\begin{aligned} A &= G\frac{i}{1-(1+i)^{-N}}\frac{(1+i)^N - Ni - 1}{i^2(1+i)^N} \\ &= G\frac{\left[(1+i)^N - 1\right] - Ni}{i\left[(1+i)^N - 1\right]}. \end{aligned}$$

Dividing each term of the numerator of this equation by the denominator yields the equation for UW_G:

$$\boxed{UW_G \equiv A = G\left[\frac{1}{i} - \frac{N}{(1+i)^N - 1}\right].} \tag{2-87}$$

Uniform Worth Factor

The **gradient series uniform worth factor** $(A/G,i,N)$ is *defined* as the normalized uniform worth of the gradient series—$UW_G G^{-1}$:

$$\boxed{(A/G,i,N) \equiv \frac{UW_G}{G} = \frac{1}{i} - \frac{N}{(1+i)^N - 1}.} \tag{2-88}$$

Hence, we calculate UW_G with the value of $(A/G,i,N)$:

$$\boxed{UW_G \equiv A = G(A/G,i,N).} \tag{2-89}$$

EXAMPLE 2.17 Converting a Complicated Deposit Series of a Savings Account to a Uniform Series of Deposits (*E*-Class)

Problem: A friend wants to start saving some of his monthly salary. On January 2nd he will deposit \$1,500 in a new savings account with an interest rate of 0.5% per month. On January 31st he will deposit \$500, and each subsequent end-of-month deposit will increase by \$100. *(1) To avoid the need to keep track of the different sizes of the 13 deposits, he asks you to create a series of 12 constant end-of-month deposits that is equivalent to them. (2) Check the result by comparing the balances on December 31st of the 13 deposit series and the equivalent 12 deposit uniform series.*

Solution: Figure 2.19 shows the CFD of the original 13 deposits. It consists of (1) a deposit of \$1,500 at the *beginning* of the first interest period, (2) a uniform series of 12 \$500 deposits that begin at the *end* of the first interest period, and (3) a gradient series

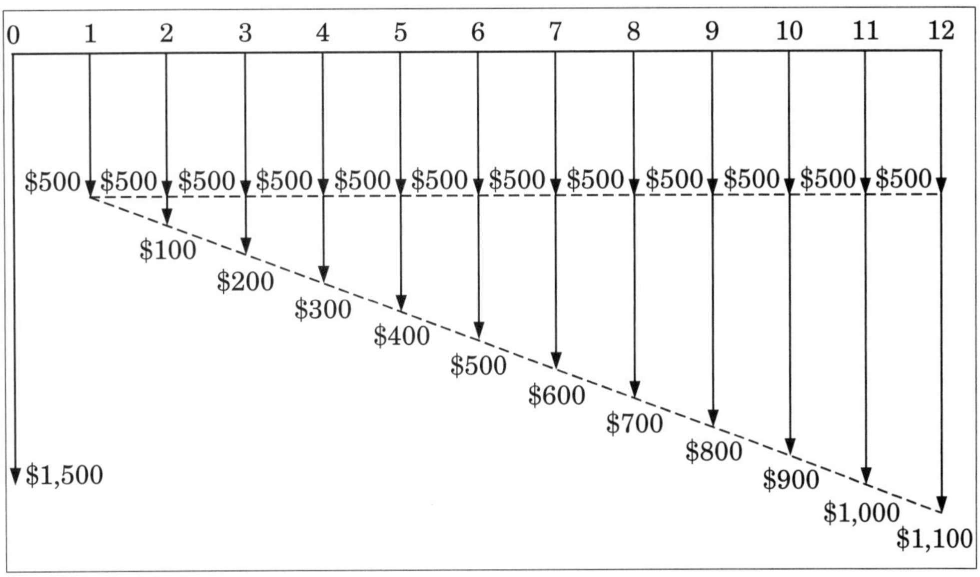

Figure 2.19 **CFD of the 13 Deposits Proposed by the Owner**

(whose metric is \$100) of 11 deposits that begin at the end of the second interest period. December 31st is the end of 12th interest period.

(1) Calculate the Metric of the Equivalent Uniform Series

To be equivalent to the original deposit series, the metric A of the uniform series of 12 deposits must equal the sum of the uniform worths of the \$1,500 deposit and the uniform (\$500) and gradient (\$100) series of deposits[12]:

$$A = \$1{,}500\underbrace{(A/P, 0.5\%, 12)}_{0.086066} + \$500 + \$100\underbrace{(A/G, 0.5\%, 12)}_{5.44057} = \underline{\underline{\$1{,}173.16}}.$$

(2) Calculate the Balances on December 31st

The account balance at the end of the 12th interest period is the future worth of the *equivalent* uniform series:

$$B_{12} = \$1{,}173.16\underbrace{(F/A, 0.5\%, 12)}_{12.33556} = \underline{\underline{\$14,471.59}}.$$

The account balance at the end of the 12th interest period is also the sum of the future worths of the *original* deposits:

$$B_{12} = \$1{,}500\underbrace{(F/P, 0.5\%, 12)}_{1.061678} + \$500\underbrace{(F/A, 0.5\%, 12)}_{12.33556} + \\ + \$100\underbrace{(F/G, 0.5\%, 12)}_{67.11247} = \underline{\underline{\$14{,}471.54}}.$$

Except for negligible factor round-off errors, the two balances are equal because A is defined as the metric of a uniform series that is equivalent to the original \$1,500 deposit, the \$500 uniform series, and the \$100 gradient series.

2.8 GEOMETRIC GRADIENT SERIES FACTORS

2.8.1 Present Worth Factor

Present Worth of the Series

The cash flows of a **geometric gradient series** *change* from period-to-period at a *fixed rate g* that can be positive or negative. When $g < 0$, the series is often called a **decreasing geometric gradient series**. Figure 2.20 is the CFD of a geometric gradient series with N cash flows. Like a uniform series, a geometric gradient series has a cash flow A_1 at the end of the first interest period.

The magnitudes of the cash flows of a geometric gradient series are *defined* by

[12] If two series are equivalent, their uniform worths on any common timespan must also be equal. The uniform worth of a uniform series is just its metric.

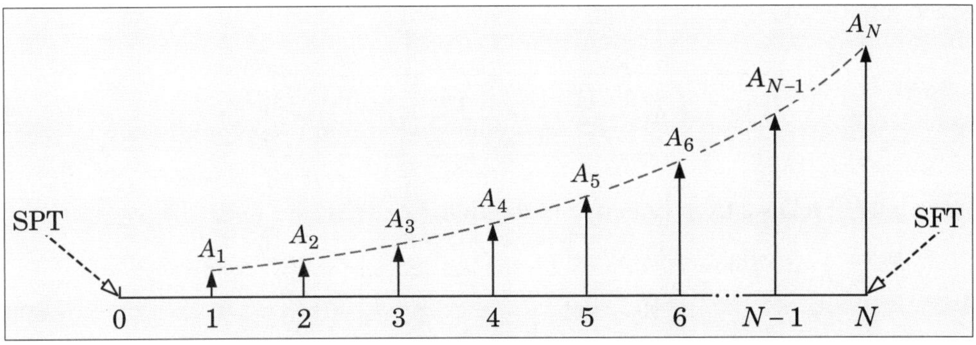

Figure 2.20 **CFD of a Geometric Gradient Series with a *Positive Value of g***

$$\boxed{A_k \equiv A_1(1+g)^{k-1} \qquad k = 1, 2, \ldots, N,} \tag{2-90}$$

where A_1 and g are the **metrics** of the series. Their values must be specified.

Practice Note: *A geometric gradient series is often used to model costs because an annual cost increase of an object, such as an automobile, a work station, or a plane, is usually stated as a percentage of its cost in the previous year.*

The **present worth of a geometric gradient series** PW_{A_1} is given by

$$\begin{aligned} PW_{A_1} &= \frac{A_1}{(1+i)} + \frac{A_1(1+g)}{(1+i)^2} + \cdots + \frac{A_1(1+g)^{N-1}}{(1+i)^N} \\ &= \sum_{k=1}^{N} \frac{A_1(1+g)^{k-1}}{(1+i)^k}, \end{aligned} \tag{2-91}$$

where the **series present time** SPT is *defined* as *one* interest period *before* the timeline point of the first cash flow of the series, and the **series future time** SFT is *defined* as the timeline point of its Nth cash flow.

Present Worth Equation

The sequence of the present worths in Equation (2-91) is a geometric sequence whose equations for a_1 and r are

$$a_1 = \frac{A_1}{1+i} \tag{2-92}$$

and

$$r \equiv \frac{a_{k+1}}{a_k} = \frac{A_1(1+g)^k}{(1+i)^{k+1}} \left[\frac{A_1(1+g)^{k-1}}{(1+i)^k} \right]^{-1} = \frac{1+g}{1+i} \qquad k = 1, 2, \ldots, N-1. \tag{2-93}$$

Hence, the sum of the present worths is given by Equation (2-44):

$$PW_{A_1} = \frac{A_1}{1+i} \frac{1-(1+i)^{-N}(1+g)^N}{1-(1+i)^{-1}(1+g)}, \tag{2-94}$$

which reduces to

$$PW_{A_1} = A_1 \frac{1-(1+i)^{-N}(1+g)^N}{i-g} \qquad g \neq i. \tag{2-95}$$

When $g = i$, it follows from Equation (2-91) that

$$PW_{A_1} = A_1 \frac{N}{1+i} \qquad g = i. \tag{2-96}$$

Factor Definition

The **geometric gradient series present worth factor** $(P/A_1, g, i, N)$ is *defined* as the normalized present worth of the series—$PW_{A_1}(A_1)^{-1}$:

$$(P/A_1, g, i, N) \equiv \frac{PW_{A_1}}{A_1} = \frac{1-(1+i)^{-N}(1+g)^N}{i-g} \qquad g \neq i. \tag{2-97}$$

Hence, we calculate PW_{A_1} with the value of $(P/A_1, g, i, N)$:

$$PW_{A_1} = A_1(P/A_1, g, i, N) \qquad g \neq i. \tag{2-98}$$

Computation Note: *The presence of the third variable g in the present worth equations of a geometric gradient series makes it* ***impractical*** *to tabulate their corresponding factor values.*

2.8.2 Net Present Worth Factor

Present Worth Equation

A geometric gradient series is often used to model the yearly maintenance costs of new equipment. However, the first year cost is usually covered by a manufacturer's warranty. Hence, the present worth of this cost must be subtracted from the present worth of the geometric gradient series of costs. When we do this, Equation (2-95) and Equation (2-96) become

$$NPW_{A_1} = A_1 \left[\frac{1-(1+i)^{-N}(1+g)^N}{i-g} - (1+i)^{-1} \right] \qquad g \neq i. \tag{2-95 net}$$

$$NPW_{A_1} = A_1 \frac{N-1}{1+i} \qquad g \neq i. \tag{2-96 net}$$

Factor Definition

The **geometric gradient series net present worth factor** $(P/A_1, g, i, N)_{\text{net}}$ is *defined* as the normalized net present worth of the series:

$$(P/A_1, g, i, N)_{\text{net}} \equiv \frac{NPW_{A_1}}{A_1} = (P/A_1, g, i, N) - \frac{1}{1+i} \qquad g \neq i. \qquad \text{(2-97 net)}$$

EXAMPLE 2.18 Using a Geometric Gradient Series to Model the Maintenance Costs of the Compressors (*PWC*-Class)

Problem: When gradient series are used to model the annual maintenance costs of the compressors in Example 2.16, the maintenance costs in the tenth year are nine times larger than those in the second year. However, a subsequent review of the maintenance cost histories of similar compressors indicates that this ninefold increase is unrealistic. It appears that a geometric gradient series with a value of g equal to 0.05 is a better model of the annual compressor maintenance costs. *Which compressor has the least-cost if the values of A_1 equal the corresponding values of G used in Example 2.16 (see Table 2.2)?* NOTE: The maintenance costs in the 1st year are covered by warranties.

Solution: We only need to calculate the ***net*** present worths of the annual maintenance costs with Equation (2-97 net) and add them to the previously calculated present worths of the *P*, *S*, and *A* cost elements in Example 2.16. This calculates NPW_{A_1} for compressor *X*:

$$NPW_{A_1,X} = \$750\left[\underbrace{(P/A_1, 5\%, 20\%, 10)}_{4.91283} - \frac{1}{1+0.2}\right] = \$3,060.$$

The value of NPW_{A_1} for compressor *Y* is calculated in the same way:

$$NPW_{A_1,Y} = \$460\left[\underbrace{(P/A_1, 5\%, 20\%, 10)}_{4.91283} - \frac{1}{1+0.2}\right] = \$1,877.$$

Finding the Least-Cost Compressor

We can now calculate PW_X and PW_Y:

$$\begin{aligned} PW_X &= P + PW_A + NPW_{A_1} - PW_S \\ &= \$80,000 + \$12,577 + \$3,060 - \$646 = \underline{\underline{\$94,991}} \\ PW_Y &= P + PW_A + NPW_{A_1} - PW_S \\ &= \$85,000 + \$9,223 + \$1,877 - \$727 = \underline{\underline{\$95,373}}. \end{aligned}$$

Now, *compressor X is the least-cost alternative because of the large reduction in the difference of the compressor maintenance costs.*

2.8.3 Future Worth Factor

Future Worth Equation

The future worth of *any* series is the product of its present worth and $(1+i)^N$. It follows from Equation (2-95) that

$$FW_{A_1} = A_1 \frac{1-(1+i)^{-N}(1+g)^N}{i-g}(1+i)^N \qquad g \neq i.$$

Hence,

$$FW_{A_1} = A_1 \frac{(1+i)^N - (1+g)^N}{i-g} \qquad g \neq i. \tag{2-99}$$

When $g = i$, it follows from Equation (2-96) that

$$FW_{A_1} = A_1 N(1+i)^{N-1} \qquad g = i. \tag{2-100}$$

Future Worth Factor

The **geometric gradient series future worth factor** $(F/A_1, g, i, N)$ is *defined* as the normalized future worth of the series—$FW_{A_1}(A_1)^{-1}$:

$$\boxed{(F/A_1, g, i, N) \equiv \frac{FW_{A_1}}{A_1} = \frac{(1+i)^{-N} - (1+g)^N}{i-g} \qquad g \neq i.} \tag{2-101}$$

Hence, we calculate FW_{A_1} with the value of $(F/A_1, g, i, N)$:

$$\boxed{FW_{A_1} = A_1(F/A_1, g, i, N) \qquad g \neq i.} \tag{2-102}$$

EXAMPLE 2.19 Another Way to Create a Retirement Series (*WC*-Class)

Problem: The engineer in Example 2.13 has decided to change her mutual fund purchasing strategy. The annual purchases of mutual fund shares will start on her 25th birthday and will increase by 10% per year; they will end on her 59th birthday. *What is the size of her first purchase if the estimated interest rates of the mutual and bond funds are still 8% and 5% per year, respectively?*

Solution: Figure 2.21 shows the revised CFD for the engineer's investment plan. The annual purchases of mutual fund shares is a geometric gradient series whose value of *g*% is 10% per year. It was shown in Example 2.13 that the principle invested in the bond fund must be $1,500,000 when the fund's interest rate is 5% per year. Hence, the future worth of the geometric gradient series of mutual fund purchases must be $1,500,000 on the engineer's 59th birthday. To calculate the size of the first purchase A_1, we rearrange Equation (2-102):

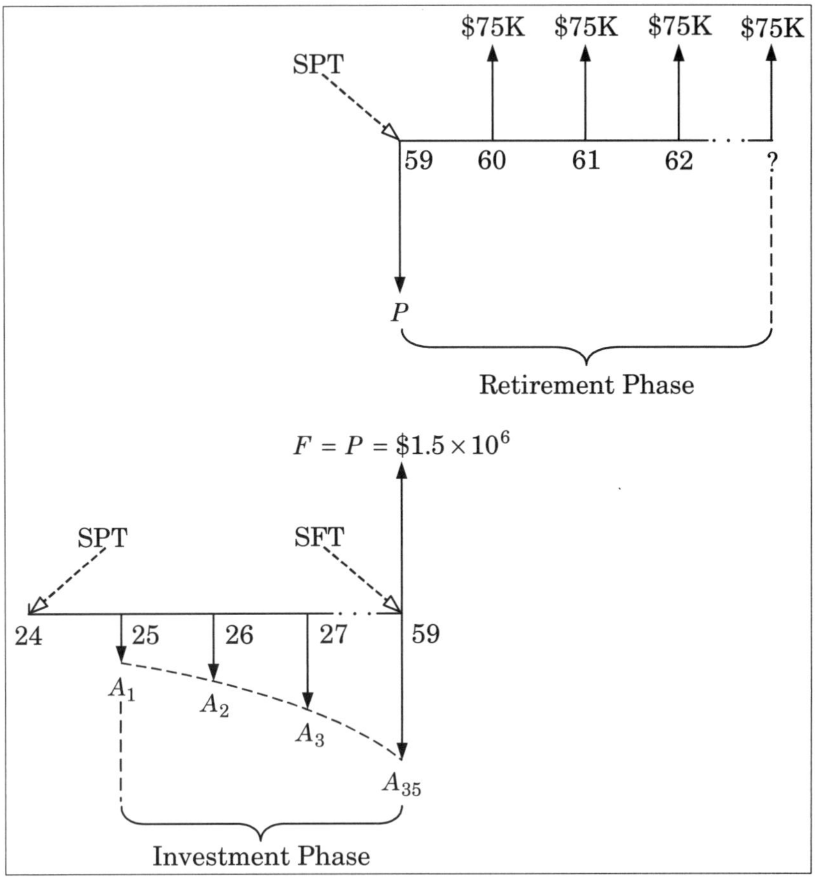

Figure 2.21 **CFD of the Engineer's Revised Investment Plan**

$$A_1 = \frac{\$1.5\times10^6}{(F/A_1, g, i, N)} = \frac{\$1.5\times10^6}{\underbrace{(F/A_1, 10\%, 8\%, 35)}_{665.85}} = \underline{\underline{\$2{,}252.76.}}$$

2.8.4 Uniform Worth of a Geometric Gradient Series

Definition

We often need to transform a geometric gradient series into an *equivalent* uniform series with the *same* SPT and number of interest periods. Figure 2.22 shows this transformation.

The **uniform worth of a geometric gradient series** UW_{A_1} is *defined* as the *metric* A of an *equivalent* uniform series. The two series are equivalent when their present worths are equal:

$$PW_A = PW_{A_1}.$$

Substituting the expressions for PW_A and PW_{A_1} from Equations (2-45) and (2-95), respectively, into this equation yields

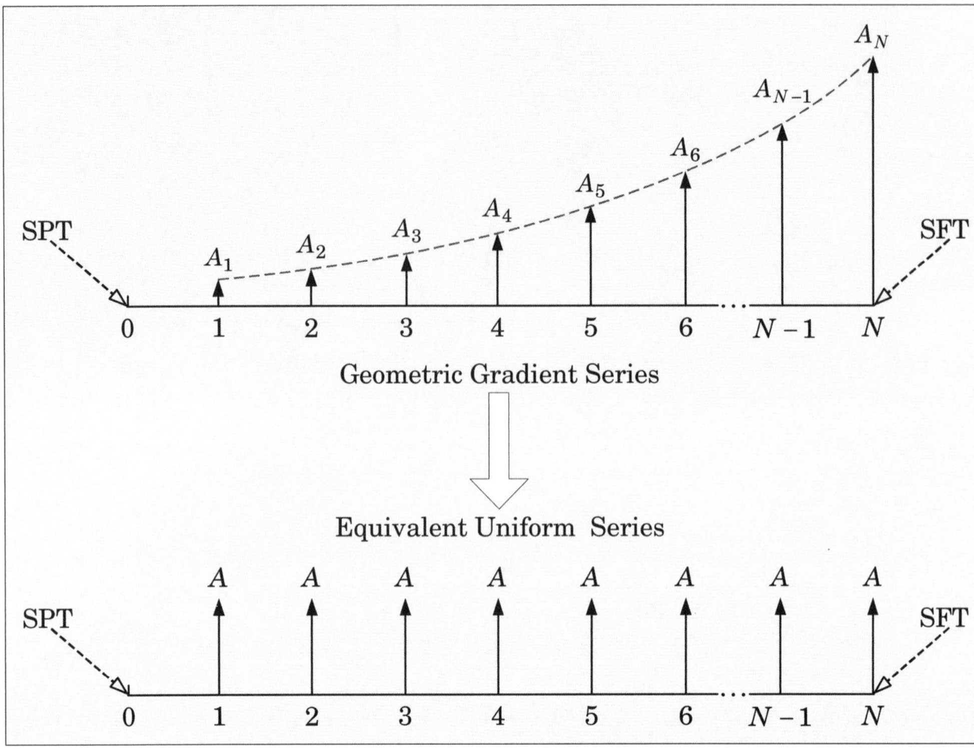

Figure 2.22 **Transforming a Geometric Gradient Series to a Uniform Series**

$$\underbrace{A\frac{1-(1+i)^{-N}}{i}}_{PW_A} = \underbrace{A_1\frac{1-(1+i)^{-N}(1+g)^N}{i-g}}_{PW_{A_1}} \qquad g \neq i.$$

Solving this equation for the metric A yields

$$\boxed{UW_{A_1} = A = A_1\frac{i}{(1+i)^N-1}\frac{(1+i)^N-(1+g)^N}{i-g} \qquad g \neq i.} \tag{2-103}$$

Uniform Worth Factor

The **geometric gradient series uniform worth factor** $(A/A_1, g, i, N)$ is defined as the normalized uniform worth of the series —$UW_{A_1}(A_1)^{-1}$:

$$\boxed{\left(A/A_1, g, i, N\right) \equiv \frac{UW_{A_1}}{A_1} = \frac{i}{(1+i)^N-1}\frac{(1+i)^N-(1+g)^N}{i-g} \qquad g \neq i.} \tag{2-104}$$

Hence, we calculate UW_{A_1} with the value of $\left(A/A_1, g, i, N\right)$:

$$\boxed{UW_{A_1} = A_1\left(A/A_1, g, i, N\right) \qquad g \neq i.} \tag{2-105}$$

2.9 RATE of RETURN of a SIMPLE INVESTMENT

2.9.1 Comparison of Loans & Investments

Definitions

An **investment** is an entity into which money—called **capital**—is transferred from an investor who *expects* the return of the capital and payment of interest over a specified time. There are **two types of investments**: business and financial. Constructing a chemical plant and selling its products and buying a pizza parlor are examples of **business investments**. Buying stocks and bonds with the expectation that their prices will rise are examples of **financial investments**.

Analogy Between Loans and Investments

Putting money into an investment is analogous to making a loan. In both cases money is placed at risk with the expectation of recovering it and obtaining a monetary reward that is commensurate with the risk. Table 2.4 lists the economic variables associated with loans and investments and compares their terminologies.

TABLE 2.4 Terminologies of Loans and Investments

ECONOMIC VARIABLE	TERMINOLOGY	
	LOAN	INVESTMENT
Money Initially at Risk	Principle	Capital
Duration of the Transaction	Term	Life
Time Interval for Earning Interest	Interest Period	Interest Period
Money Owed at the End of an Interest Period	Balance	Balance
Money Received at the End of an Interest Period	Payment	Receipt
Money Disbursed at the End of an Interest Period	Principle	Disbursement
Profitability of the Transaction	Interest Rate	Rate of Return

Uncertainty and Risk: Hallmarks of a Business Investment

The basic difference between a loan and a business investment is the certainty associated with a loan. The principle series, term, interest rate, and payment series of a loan are precisely specified. The risk of a **secured loan**—a loan with collateral, such as a mortgage on a house or a lien on an automobile—is the small probability that the borrower will not repay it.

Every economic variable in a business investment that requires construction of a plant to produce a product is uncertain!

- The ***fixed capital*** *FC* required to construct the plant is *estimated*.
- The *investment life* is *assumed*.
- The *estimates* of the *receipt* and *disbursement series* of the plant when it is producing its product(s) are based on a *forecast* of economic conditions far into the future.[13]

Hence, the *predicted* rate of return is based on an *estimate*, an *assumption*, and a *forecast* for the corresponding **key economic variables**: *fixed capital series*,

[13] The uncertainty of a financial investment is the future selling price of the stock or bond.

investment life, and receipt and disbursement series. These uncertainties make many business investments high-risk ventures. If the *actual* values of the key economic variables are *disappointing*, the investment's rate of return will also be *disappointing*.

Practice Note: *To compensate for the uncertainty and, hence, risk of a business investment, a company often requires a predicted rate of return that is five or more times greater than the interest rate of a secure loan. For example, in the chemical industry the predicted rate of return of a proposed production facility must usually be 20% or more for it to be considered for funding.*

2.9.2 Rate of Return of an Investment

Definitions

Previously the GRE

$$\sum_{k=0}^{N} \frac{R_k}{(1+i)^k} = \sum_{k=0}^{N} \frac{P_k}{(1+i)^k}$$

$$\sum_{k=0}^{N} \frac{R_k - P_k}{(1+i)^k} = 0 \tag{2-106}$$

was used in Example 2.7 to calculate the value of R_3.

We now use the GRE to find the equation for an investment's **rate of return** (ROR), which is denoted by i^* to distinguish it from the *specified* interest rate of a loan:

$$\sum_{k=0}^{N} \frac{R_k - D_k}{(1+i^*)^k} = \sum_{k=0}^{N} \frac{CF_k}{(1+i^*)^k} = 0, \tag{2-107}$$

where R_k and D_k are the **total receipts** and **disbursements**, respectively, transferred in the kth interest period.[14] Their difference is the **cash flow** CF_k of the kth interest period:

$$CF_k \equiv R_k - D_k. \tag{2-108}$$

Unlike R_k and D_k, CF_k is an *algebraic* quantity. Equation (2-107) says that ***the rate of return of an investment is the value of i* that makes the present worth of its cash flows equal zero.***

Equation (2-107) is based on a *tacit* assumption—called the **End-of-Year Convention**—that R_k, D_k, and, hence, CF_k occur at the *end* of the kth interest period, which is normally a year. However, the purchase of land or an existing plant occurs at the *beginning* of the 1st interest period, which is TP_0.

The validity of the End-of-Year Convention, which is normally used in the literature and practice of engineering economics, is discussed and illustrated in Chapter 6.

General Polynomial Equation

To obtain a polynomial equation for calculating i^*, we first multiply Equation (2-107) by $(1+i^*)^N$:

[14] Capital disbursements normally occur without any corresponding receipts.

$$CF_0(1+i^*)^N + CF_1(1+i^*)^{N-1} + \cdots + CF_{N-1}(1+i^*) + CF_N = 0. \tag{2-109}$$

Substituting the *definition*

$$x \equiv 1 + i^* \tag{2-110}$$

into Equation (2-109) yields the **General Polynomial Equation** (GPE):

$$\boxed{CF_0 x^N + CF_1 x^{N-1} + \cdots + CF_{N-1}x + CF_N = 0.} \tag{2-111}$$

Since an Nth degree polynomial with real coefficients can have a combination of positive, negative, and complex roots, we find the solution of a **one disbursement-one receipt** ($1D$-$1R$) investment to determine if negative rates of return are possible.

Interval of the Rate of Return

For a $1D$-$1R$ investment, the GPE reduces to

$$CF_0 x^N + CF_N = 0. \tag{2-112}$$

Solving this equation for x yields

$$x = \sqrt[N]{-\frac{CF_N}{CF_0}} = 0. \tag{2-113}$$

where CF_0 is negative. Replacing x in this equation by its definition $1 + i^*$ yields a simple expression for i^*:

$$i^* = \sqrt[N]{-\frac{CF_N}{CF_0}} - 1. \tag{2-114}$$

It follows from this result that

1. When $CF_N > -CF_0$, the capital is recovered, and interest is earned. Hence, $0 < i^* < \infty$.
2. When $CF_N = -CF_0$, the capital is recovered, but interest is not is earned. Hence, $i^* = 0$.
3. When $CF_N < -CF_0$, *part* of the capital is recovered. Hence, $-1 < i^* < 0$.

Thus, the value of i^* is in the interval

$$-1 < i^* < \infty. \tag{2-115}$$

Computation Note: *Since $x = 1 + i^*$, it follows Equation (2-115) that a solution of the GPE that is financially significant must be in the interval*

$$0 < x < \infty. \tag{2-116}$$

EXAMPLE 2.20 Rate of Return of a $1D$-$1R$ Financial Investment (*SI*-Class)

Problem: On September 1st, 2004, an investor bought 100 shares of Google stock for \$10,270. Two years later he sold them for \$38,099. *What is the rate of return of this investment?*

Solution: This is a *1D-1R* investment with two interest periods. It follows from the problem statement that $CF_0 = -\$10{,}270$ and $CF_2 = \$38{,}099$. Substituting these values into Equation (2-114) yields the rate of return:

$$i^* = \sqrt{-\frac{CF_2}{CF_0}} - 1 = \sqrt{-\frac{\$38{,}099}{-\$10{,}270}} - 1 = \underline{\underline{0.9261 \text{ per year}}}; \qquad i^*\% = \underline{\underline{92.61\% \text{ per year.}}}$$

This is a remarkable rate of return. The *long-term* rate of return of the stock market is *about* 10% per year.

Next, we determine a *sufficient* condition for the existence of a unique positive solution of the GPE and, hence, a unique of rate of return.

2.9.3 Simple Investments

Definition

A **simple investment** has one or more initial negative cash flows followed by one or more positive cash flows. Figure 2.23 is an example of the CFD of a simple investment. *The distinguishing property of a simple investment is the occurrence of only **one** change in the signs of its cash flows and, hence, the coefficients of its GPE.*

Practice Note: *Cash flows normally occur at the ends of **every** interest period in an investment's life. Figure 2.23 illustrates an exception to this generalization.*

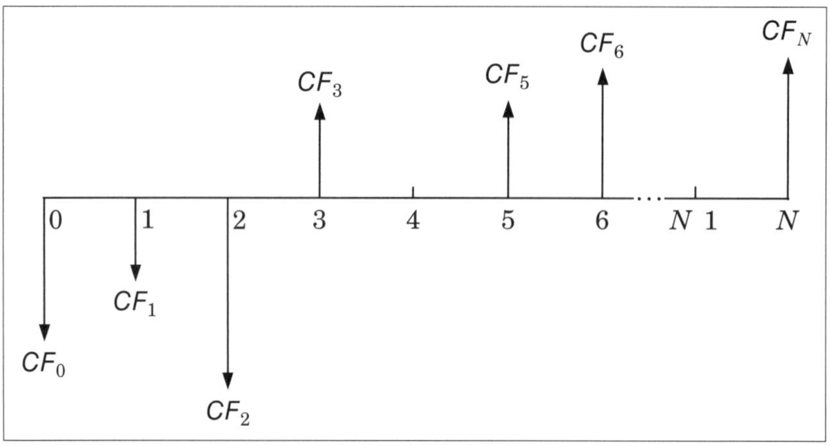

Figure 2.23 **CFD of a Simple Investment** NOTE: After the last negative cash flow CF_2, the cash flows are positive.

Uniqueness of the ROR and Positive Balances

Descartes' Rule of Signs[15] gives the number of positive roots of a polynomial: *The number of positive roots of a polynomial with real coefficients equals the number of sign changes of its coefficients or differs from it by an even integer.*

A simple investment has a unique ROR because it has only one change in the signs of its cash flows. This unique ROR is called the **internal rate of return** (IRR) to emphasize that it is generated by the investment and is not specified like the interest rate of a loan. Since most investments are simple investments, this is an extremely important result. NOTE: IRR% means percentage IRR.

[15] Check your algebra text or the Internet.

Except for B_N, which is defined as zero, the balances of a simple investment are always positive. *Subsection 6.1.2 presents the proof.*

Calculating the IRR of a Simple Investment

It's very easy to calculate the IRR of a simple investment with three *consecutive* cash flows because its GPE is a quadratic equation:

$$CF_0x^2 + CF_1x + CF_2 = 0, \tag{2-117}$$

where $CF_0 < 0$, $CF_2 > 0$, and CF_1 can be negative or positive. In either case, there is only one change in the signs of the cash flows.

The solution of Equation (2-117) is given by the quadratic formula:

$$x = \frac{-CF_1 \pm \sqrt{CF_1^2 - 4CF_0CF_2}}{2CF_0}. \tag{2-118}$$

EXAMPLE 2.21 Calculating the IRR% Analytically (*SI*-Class)

Problem: *Calculate the IRR% of an investment with the following cash flows:*

Cash Flow, $	–3,000	2,200	1,000
Interest Period *k*	0	1	2

Solution: The roots of a quadratic GPE are given by Equation (2-118). Since there is only one change in the signs of the three consecutive cash flows, there is a unique positive root:

$$x = \frac{-CF_1 \pm \sqrt{CF_1^2 - 4CF_0CF_2}}{2CF_0} = \frac{-\$2,200 \pm \sqrt{-\$2,200^2 - 4(-\$3,000)\$1,600}}{2(-\$3,000)}$$
$$= -0.31728 \text{ and } 1.05061.$$

It follows from Equation (2-110) that

$$i^* = x_+ - 1 = 1.05061 - 1 = 0.05061 \text{ per year}$$
$$i^*\% = \underline{\underline{5.06\% \text{ per year.}}}$$

2.10 CALCULATING the IRR% with EXCEL

2.10.1 IRR Function

Unlike a second degree polynomial, third and fourth degree polynomials have cumbersome equations for their roots. However, we can find them and the roots of higher degree polynomials with mathematics software, such as *Mathematica*, Mathcad, and MATLAB. We will do this later in the book. But for *simple* investments, which constitute the large majority of *all* investments, we can use Excel's[16] **IRR function** to calculate the IRR%. Its syntax is given in Table 2.5:

[16]"Excel" is a trademark ® of Microsoft, Inc.

TABLE 2.5 Syntax of Excel's IRR Function

IRR(*array, guess*)
(1) The argument *array* is a column or row of cells, such as B1:B10 or A3:F3, containing a sequence (first to last) of one or more *negative* cash flows followed by one or more *positive* cash flows
•If the array contains some text and empty cells, Excel ignores them.
(2) The argument *guess* is an estimate of the IRR% that Excel uses to start its iterative procedure for calculating it. Normally, you don't have to provide *guess*. If it is omitted, Excel assumes that *guess* equals 10%.
•If the IRR% is negative, you will usually need to provide a value for *guess*.
•If Excel can't converge to the IRR% after 20 iterations, it returns # NUM!. Hence, you must provide a better value for *guess* to obtain the IRR%.

EXAMPLE 2.22 Calculating the IRR% of an Investment with Excel (*SI*-Class)

Problem: *Calculate the IRR% of an investment with the following cash flows:*

CASH FLOW, $$10^6$	−10	−11	3	4	8	0	15	9	8	4	2
YEAR *k*	0	1	2	3	4	5	6	7	8	9	10

Solution: Since there is only one change in the signs of the cash flows, we can use Excel's IRR function to calculate the IRR% of this *simple* investment. Spreadsheet 2.2 shows that the IRR% is 19.46% per year. Subsection 2.10.2 explains the three types of spreadsheet cells.

	A	B	C	D
1				
2		***IRR% of a Simple Investment***		
3		***k***	**Cash Flow, $$10^6$**	
4		0	(10)	
5		1	(11)	
6		2	3	
7		3	4	
8		4	8	
9		5	0	
10		6	15	
11		7	9	
12		8	8	
13		9	4	
14		10	2	
15			19.46%	**=IRR(C4:C14)**

Spreadsheet 2.2 NOTE: Label cell D15 explains the number in formula cell C15.

2.10.2 Spreadsheet Cells

Types

Spreadsheet 2.2 displays the **three types** of **spreadsheet cells**: input, formula, and label.

1. An **input cell** contains the value of an input variable of a problem. If it is *formatted* (displayed) as a percentage or currency, Excel converts it to a number in evaluations. A *negative* number can be displayed as the magnitude of the number enclosed in parentheses, such as (10) in cell C4 of Spreadsheet 2.2. This is useful because the minus sign of a text font is relatively small: -10. The cells in array B4:C14 are input cells.
2. A **formula cell** contains an expression (Microsoft calls it a formula.) that is evaluated, and the result is displayed in the cell. Cell C15 is a formula cell. NOTE: A formula must be preceded by an equal character (=). In the Excel calculations in this book, a formula is written in terms of the cell addresses of the input variables and one of Excel's computational functions, such as IRR, PV, and SUM.[17]
3. **Label cells** are *shaded* cells that contain *explanatory* material that begins with a single quote character (') that is not displayed. They explain the meaning of the numbers in adjacent input and formula cells. For example, cells B3 and C3 explain the meanings of the numbers in arrays B4:B14 and C4:C14, respectively. Cell D15 explains the number in formula cell C15. A label cell can also contain an inserted Excel drawing object, such as a brace ({) that spans an array.

Book Conventions: *(1) A negative number is displayed as the magnitude of the number enclosed in parentheses. See array C4:C5 in Spreadsheet 2.2.*
(2) Label cells are shaded. See array B3:C3 and cell D15.
(3) The result of a spreadsheet is displayed with a bold type, and its cell has a thick black border. An adjacent label cell shows the formula (with a bold type) that calculated the result. See cell D15.

Practice Note *A spreadsheet with sparse annotations for its input variables and the Excel formula(s) used to solve the spreadsheet's problem is **worthless**. A reader will not understand the spreadsheet, and its author will soon forget its meaning.*

Cell Shading

(1) To shade several cells, *Select* the cells > *Right Click* the last selected cell > *Click* **Format Cells** > *Click* the **Fill** tab > *Click* a light grey button > *Click* **OK.**
(2) To shade several cells with an *existing* cell's shade, *Select* a shaded cell > *Double Click* the **Format Painter** button (a yellow paint brush on the left side of the **Home** tab) > *Click* the cells to be shaded > *Press* **Enter.**

2.10.3 Goal Seek Command

Sometimes we want to know the value of a spreadsheet input variable that will yield a desired result. For example, what value of the cash flow in cell C4 of Spreadsheet 2.2 will yield an IRR% of 20.00%? To find the answer, we use the Excel **Goal Seek** command explained in Step-by-Step 2.1. Spreadsheet 2.3 shows that when the number in cell C4 is (9.58), the IRR% is 20.00% per year.

[17] You can check Excel's help for a broader definition of a formula and what it can do.

Using the Excel Goal Seek Command: Step-by-Step 2.1
(1) *Click* the **Data** tab. In the **Data Tools** group *Click* **What-If Analysis** > *Click* **Goal Seek...** to open the **Goal Seek** dialog box.
(2) In the **Set cell:** text box *Click* the cell that contains the value you want to change. (To change the IRR% in Spreadsheet 2.2, *Click* cell C15.)
(3) In the **To value:** text box *Type* the desired value. (To change the IRR% in Spreadsheet 2.2 to 20%, *Type* 20.00%.)
(4) In the **By changing cell:** text box *Click* the cell that contains the value you want Excel to change (To change the value in cell C4 of Spreadsheet 2.2, *Click* cell C4.) > *Click* **OK**.
(5) After the iteration converges, *Click* **OK** in the **Goal Seek** status box.

	A	B	C	D
1				
2		***IRR% of a Simple Investment***		
3		***k***	**Cash Flow, $$10^6$**	
4		0	(9.58)	
5		1	(11)	
6		2	3	
7		3	4	
8		4	8	
9		5	0	
10		6	15	
11		7	9	
12		8	8	
13		9	4	
14		10	2	
15			**20.00%**	**=IRR(C4:C14)**

Spreadsheet 2.3 NOTE: Goal Seek changed the number in cell C4 of Spreadsheet 2.2 to (9.58) to obtain an IRR% of 20.00% per year.

⚠ ***The original number (10) in cell C4 was formatted with zero decimal places. If the decimal formatting is not changed, the goal seek result will also be 10 because 9.58 formatted with zero decimal places is 10.***

2.11 CLASSIFICATION of the EXAMPLES

There are 22 examples in Chapter 2. *They illustrate four of the nine major classes of problems in engineering economics listed in Table 2.6.* Hence, if you understand how to solve these four classes of problems, you understand the contents of this chapter on the foundation of engineering economics. The remaining five classes of problems are discussed in subsequent chapters.

TABLE 2.6 The 10 Major Classes of Problems in Engineering Economics

PROBLEM CLASS		INTEREST RATE	CHAPTER EXAMPLES
#	NAME		
1	Equivalence (*E*-Class)	Equal	2.1, 2.2, 2.3, 2.4, 2.6, 2.7, 2.8, 2.10, 2.11, 2.12, 2.14, 2.15, 2.17
2	Present Worth Comparison (*PWC*-Class)	Equal	2.9, 2.16, 2.18 and Chapter 5
3	Annual Worth Comparison (*AWC*-Class)	Equal	Chapter 4
4	Worth Equality (*WE*-Class)	Different	2.13, 2.19
5	Simple Investment (*SI*-Class)	Unknown	2.5, 2.20, 2.21, 2.22 and Chapter 6
6	Quasi-Simple Investment (*QSI*-Class)	Unknown	Chapter 6
7	Non-Simple Investment (*NSI*-Class)	Unknown	Chapter 7
8	Mutually Exclusive Investments (*MEI*-Class)	Equal	Chapter 8
9	Income Taxes (*IT*-Class)	Unknown	Chapter 9
10	Sensitivity Analysis (*SA*-Class)	Known	Chapter 10

2.12 DEVELOPING the *12-FACTOR Calculator*

2.12.1 Creating the 12-Factor Functions in Excel

To avoid calculating values of the three nontabulated factors of a geometric gradient series and nontabulated values of the nine tabulated factors, I used the Visual Basic Editor in Excel to *define* 12 functions that represent the 12-Factors. However, in Excel we can't define a factor with the conventional mnemonic function notation. Hence, the mnemonic notation shown in Table 2.7 was used to define the 12-Factors.

TABLE 2.7 Definitions of the 12-Factors in Excel

PRESENT WORTH	UNIFORM WORTH	FUTURE WORTH
P_F$(i, N) \equiv (P/F, i, N)$	A_P$(i, N) \equiv (A/P, i, N)$	F_P$(i, N) \equiv (F/P, i, N)$
P_A$(i, N) \equiv (P/A, i, N)$	A_F$(i, N) \equiv (A/F, i, N)$	F_A$(i, N) \equiv (F/A, i, N)$
P_G$(i, N) \equiv (P/G, i, N)$	A_G$(i, N) \equiv (A/G, i, N)$	F_G$(i, N) \equiv (F/G, i, N)$
P_A1$(g, i, N) \equiv (P/A_1, g, i, N)$	A_A1$(g, i, N) \equiv (A/A_1, g, i, N)$	F_A1$(g, i, N) \equiv (F/A_1, g, i, N)$

2.12.2 12-FACTOR Calculator

Definition

The ***12-FACTOR Calculator*** is displayed in Spreadsheet 2.4. It calculates the values of the 12-Factors for *arbitrary* values of $i\%$, N, and $g\%$ when $g\% \neq i\%$. Since some problems require factors with several values of N, the spreadsheet provides six *input* cells for them.

	A	B	C	D	E	F	G	H
1								
2		***12-FACTOR Calculator***						
3				***i%***	***g%***			
4			***Input:***	6.0000%	4.0000%			
5			***Number of Interest Periods N***					
6		***Input:***	2	4	6	8	10	12
7			***Solution: Present Worth Factor Values***					
8		$(P/F, i, N)$	0.89000	0.79209	0.70496	0.62741	0.55839	0.49697
9		$(P/A, i, N)$	1.83339	3.46511	4.91732	6.20979	7.36009	8.38384
10		$(P/G, i, N)$	0.89000	4.94552	11.45935	19.84158	29.60232	40.33686
11		$(P/A_1, g, i, N)$	1.86899	3.66812	5.40000	7.06714	8.67197	10.21680
12			***Solution: Annual Worth Factor Values***					
13		$(A/P, i, N)$	0.54544	0.28859	0.20336	0.16104	0.13587	0.11928
14		$(A/F, i, N)$	0.485437	0.22859	0.14336	0.10104	0.07587	0.05928
15		$(A/G, i, N)$	0.48544	1.42723	2.33040	3.19521	4.02201	4.81126
16		$(A/A_1, g, i, N)$	1.01942	1.05859	1.09816	1.13806	1.17824	1.21863
17			***Solution: Future Worth Factor Values***					
18		$(F/P, i, N)$	1.12360	1.26248	1.41852	1.59385	1.79085	2.01220
19		$(F/A, i, N)$	2.06000	4.37462	6.97532	9.89747	13.18079	16.86994
20		$(F/G, i, N)$	1.00000	6.24360	16.25531	31.62447	53.01325	81.16569
21		$(F/A_1, g, i, N)$	2.10000	4.63092	7.66000	11.26395	15.53017	20.55821

Spreadsheet 2.4 ***12-FACTOR Calculator***

Usage Notes

(1) You enter **different** values of $i\%$ and $g\%$ in input cells D4 and E4 and six values of N in input array C6:H6. **All input cells have bold borders.**

(2) If you leave out an N value, Excel assumes it is 0 and returns error messages (#VALUE!) in the corresponding uniform worth formula cells. This occurs because the denominators of the four uniform worth factors are all 0 when N equals 0.

(3) The values of the 24 present worths, 24 uniform worths, and 24 future worths are calculated in the formula arrays C8:H11, C13:H16, and C18:H21, respectively.

(4) Every cell except the eight input cells is **locked** to prevent the accidental deletion or modification of its contents.

(5) If you inadvertently assign the same values to $g\%$ and $i\%$, Excel returns an error message (#VALUE!) for each factor value of the geometric gradient series. This occurs because the denominators of the present, uniform, and future worth factors of the geometric gradient series are all 0 when $g\% = i\%$.

(6) *Values of the 12-Factors are available in any spreadsheet of a file that contains the 12-FACTOR Calculator.* For example, typing =F_P(20%, 10) in any cell in any such spreadsheet and pressing Enter returns 6.19174.

Computation Note: *When $g\% = i\%$, we use the equations derived for the worths of a geometric gradient series. Hence, to calculate PW_{A_1} and FW_{A_1}, we use Equations (2-96) and (2-100), respectively.*

Formatting the Input & Formula Cells

The input and formula cells are formatted as follows:

1. ***The cells containing i% (D4) and g% (E4) are formatted as percentages with four displayed decimal places.*** For example, if $i\% = 6.5\%$, just type 6.5% in cell D4. It will display as 6.5000%. When you enter a value of *i%* or *g%* that is less than 1 and has five or more decimal places, it will *display* with only four. For example, entering 0.012345% will display as 0.0123%. But Excel calculates with the value 0.012345%.
2. ***The 72 cells containing the factor formulas are formatted as numbers with five displayed decimal places.*** To change this, *Right Click* a cell>*Click* **Format Cells**>*Click* the **Number** tab>*Click* **Number** in the **Category:** list box>*Change* the number in the **Decimal Places:** list box>*Click* **OK**.
3. ***If any formula cell displays a string of number characters, such as ####, it means the cell's width is too small to display the number.*** Just *increase the cell's column width to display it*. To do so, *Click* the border on the RHS of the cell's column heading and drag it to the right to increase the column width.

2.12.3 Excel's Three Factor Functions

Excel has three functions that can be used as six factors. Table 2.8 compares their notations with those of the corresponding conventional factors and the ***12-FACTOR Calculator***. As you can see, the Excel functions have three disadvantages:

1. Each function represents *two* factors.
2. The arguments of the functions are more than *i%* and *N*.
3. The functions are not mnemonic.

Book Convention: *(1) The factor values used in the book were easily and quickly obtained with the* ***12-FACTOR Calculator****. Using it eliminates the inherent limitations of factor tables and Excel's limited factor functions.*
(2) However, conventional factor notation is used to explain calculations.

TABLE 2.8 The Three Factor Notations

CONVENTIONAL FACTOR	*12-FACTOR Calculator*	EXCEL FACTOR
$(P/F, i, N)$	P_F(i, N)	PV(i, N, 0, −1)

TABLE 2.8 The Three Factor Notations *(continued)*

$(P/A,i,N)$	P_A(i,N)	PV($i,N,-1$)
$(A/P,i,N)$	A_P(i,N)	PMT($i,N,-1$)
$(A/F,i,N)$	A_F(i,N)	PMT($i,N,0,-1$)
$(F/P,i,N)$	F_P(i,N)	FV($i,N,0,-1$)
$(F/A,i,N)$	F_A(i,N)	FV($i,N,-1$)

2.13 SUMMARY

The elements of a loan are defined in Section 2.1, and a *cash flow* and *cash flow diagram* are explained in Section 2.2. The **Fundamental Repayment Equation** (FRE), Equation (2-14),

$$\boxed{F = P(1+i)^N}$$

is derived in Subsection 2.3.2, and five examples illustrate its use.

The *future* and *present worths* of a *cash flow* are defined in Subsection 2.4.1, and Example 2.6 illustrates the calculation of the present worth of a bond. The *future worth factor* and *present worth factor* of a cash flow are defined in Subsection 2.4.2, and it shows how they are used to calculate the future worth and present worth of a cash flow. Subsection 2.4.2 also shows part of a page from factor tables that list the values of nine factors for various values of the interest rate and number of interest periods.

The ***12-FACTOR Calculator*** spreadsheet in Subsection 2.4.3 calculates *the values of the 12 factors of engineering economics for* ***arbitrary*** *values of i%, N, and g% ≠ i%.*

The present and future worths of a series are defined in Subsection 2.4.4, and ***equivalence*** is ***defined*** in Subsection 2.4.5. It shows that *if two series are equivalent at one timeline point, they are equivalent at all timeline points.* In addition, the interest rate for equivalent series is *unique*.

The Balance Equation is defined in Section 2.5, and it is used to derive the **General Repayment Equation** (GRE), Equation (2-37), for a loan:

$$\boxed{\sum_{k=0}^{N} \frac{R_k}{(1+i)^k} = \sum_{k=0}^{N} \frac{P_k}{(1+i)^k}.}$$

This equation says that *the payment and principle series of a loan are equivalent.*

Sections 2.6, 2.7, and 2.8 derive the present and future worth factors of a uniform series, gradient series, and geometric gradient series, respectively. Subsections 2.6.3 and 2.6.4 derive the capital recovery and sinking fund factors, respectively, of a uniform series. Subsection 2.7.3 derives the uniform worth factor for converting a gradient series to an *equivalent* uniform series. Subsection 2.8.4 does the same thing for a geometric gradient series.

Example 2.16 uses the present worth factors for a uniform series and a gradient series as part of a **present worth cost comparison** of two compressors. Example 2.18 does the same thing by replacing the gradient series with a geometric gradient series.

Subsection 2.9.2 defines the rate of return of an investment and derives the **General Polynomial Equation** (GPE), Equation (2-111), for finding it:

$$\boxed{CF_0x^N + CF_1x^{N-1} + \cdots + CF_{N-1}x + CF_N = 0.}$$

Subsection 2.9.3 defines a *simple* investment and shows that the GPE for it has only one positive root and, hence, a unique rate of return.

Subsection 2.10.1 shows how to calculate the rate of return of a simple investment with Excel's IRR function, and Subsection 2.10.3 explains how to use Excel's Goal Seek command to change the value of an investment variable to obtain a *specified* IRR%. Table 2.6 in Section 2.11 classifies the examples in Chapter 2 according to problem class.

Section 2.12 describes the development of the ***12-FACTOR Calculator***.

2.14 PROBLEMS

(P2-1) In 1626 the Lenape Native Americans sold the island of Manhattan to a Dutch trading company for $24 worth of trade goods. **(Part A)** *If the Native Americans had accepted $24 instead and deposited it on January 1st in a bank account with an interest rate of 4% per year, what was the account balance at the end of 2012?* **(Part B)** *If the bank account had an APR of 4% compounded daily, what was the account balance at the same time?*

(P2-2) *What is the opening deposit of a bank account that will provide a 5-year balance of $15,000 if the account APR is 5% compounded quarterly?*

(P2-3) *What is the total interest earned by the bank account in* **(P2-2)**?

(P2-4) A bank account with an interest rate of 6% per year has an initial deposit P_0 and another one at the end of the third year that is three times as large. There is a series of three annual withdrawals beginning with $10,000 at the end of the fourth year and increasing by $500 per year in the next two years. The last withdrawl closes the account. **(Part A)** *What is the total principle P_T of the account?* **(Part B)** *Calculate the future worths of the principles and withdrawals at the end of 10 years. Explain your result.*

(P2-5) *Derive the equation for UW_{A_1} when $g\% = i\%$. Hint*: Check the derivation of UW_{A_1} when $g\% \neq i\%$.

(P2-6) A young father plans to invest in a conservative mutual fund that has a *predicted* interest rate of 8% per year to provide for his infant daughter's college education. His first purchase of fund shares will start now and will increase by 5% per year. One year *after* the 16th purchase, he will withdraw the investment's balance and deposit the money in a bank account paying 3% per year. Four constant annual withdrawals starting one year *later* will provide the four *anticipated* tuition payments of \$40,000 per year with the last one closing the account. *Sketch the CFD and determine the size of the first mutual fund purchase.*

(P2-7) A series called S has the following five cash flows:

CASH FLOW, $\$10^3$	−2	−4	3	4	8
YEAR k	0	1	2	3	4

Convert the S series to a uniform series of six cash flows with the first one at the end of the third interest period. The interest rate is 4% per year.

(P2-8) (Part A) Figure 2.24 displays the transformation of a geometric gradient series with the metrics $A_1 = \$100$ and $g = 0.04$ into an *equivalent* gradient series. *What is the metric of the gradient series?* **(Part B)** *Confirm your solution.* The interest rate is 6% per period.

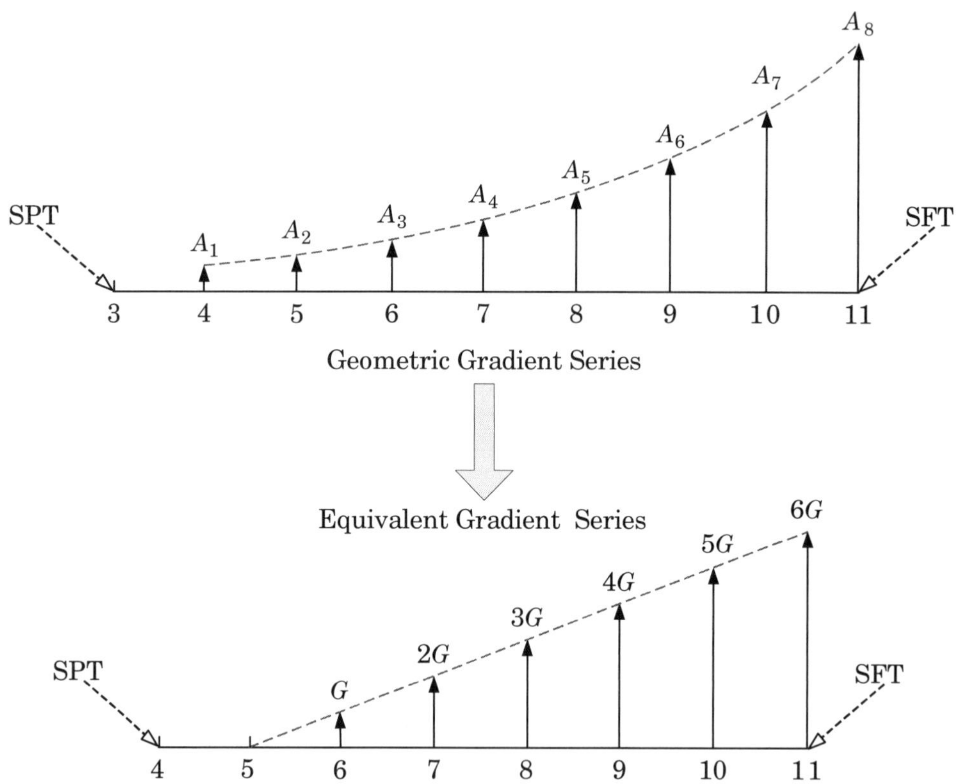

Figure 2.24 **CFD for Problem (P2-8)**

(P2-9) A winner of a million dollar lottery will receive \$50,000 per year for 20 years starting in 12 months. *How much should an investor offer the winner now to obtain an IRR% of 12.00% per year? Explain the solution procedure and display the value of the IRR% with two significant decimal digits.*

(P2-10) The *estimated* cash flows of a small personal investment are shown below:

CF, $\$10^3$	−2	−0.5	−3	0	6	0	2
YEAR k	0	1	2	3	4	5	6

(Part A) *Calculate the investment's anticipated IRR% with two significant decimal digits.* **(Part B)** *Calculate the future worths of the capital and receipt series at the end of the 10th year. Explain the result.* **(Part C)** *If the IRR% is less than 20%, what value of CF_2 will yield an IRR% of 20%. Explain how you calculated this result.*

(P2-11) Derive the following factor relationships:

$$(A/P,i,N) = (A/F,i,N) + i \tag{2-119}$$

$$(A/G,i,N) = \frac{1}{i} - \frac{N}{i}(A/F,i,N) \tag{2-120}$$

$$(F/G,i,N) = \frac{1}{i}\left[(F/A,i,N) - N\right]. \tag{2-121}$$

(P2-12) *Classify Problems (P2-1) through (P2-11).*

Computation Notes:*(1) If an equation has a factor, show it with the factor value beneath it.*
(2) Sketch a CFD when it is required to illustrate or solve the problem.
(3) Your work must clearly show what you are doing.

3

Intermediate Balances and Other Interest Rates & Series

OBJECTIVES

This chapter builds upon the foundation of engineering economics and shows you how to make some additional calculations. After reading it you will be able to

1. Calculate the intermediate balances of a loan or simple investment recursively and with the two balance equations, ***B*-Class** problems.
2. Use a ***MORTGAGE BALANCES & INTERESTS Calculator*** to determine the annual balances and interests of a mortgage, a ***B*-Class** problem.
3. Calculate the IRR% and annual percentage yield *APY* of a spec house.
4. Understand simple interest and effective annual interest rates with discrete and continuous compounding.
5. Recognize composite series and calculate their worths.

3.1 INTERMEDIATE BALANCES

3.1.1 Intermediate Balance Equations

Intermediate Balance Condition

An **intermediate balance** of a loan or a simple investment is any balance except B_0 and B_N. B_0 is always *positive*, and B_N is *defined* as 0. The repayment plan of a loan is always structured subject to the **Intermediate Balance Condition**:

Intermediate Balance Condition
The initial and intermediate balances of any loan must be positive.

The intermediate balances of simple investments always satisfy it (see Subsection 6.1.2): If this was not so, a loan would be paid off when a negative balance occurred. At that time *the lender would owe the borrower money!*

We use an **Intermediate Balance Equation** (IBE) to calculate an intermediate balance of a loan or a simple investment, a ***B*-Class** problem.

Future Worth Form of the IBE

To obtain an equation for the intermediate balance B_{k*} at any **intermediate point** k^*, we start with the form of the Balance Equation given by Equation (2-35):

$$\frac{R_k}{(1+i)^k} - \frac{P_k}{(1+i)^k} = \frac{B_{k-1}}{(1+i)^{k-1}} - \frac{B_k}{(1+i)^k} \qquad k = 1, 2, \ldots, N. \tag{3-1}$$

Summing Equation (3-1) from 1 to k^* yields

$$\sum_{k=1}^{k^*} \frac{R_k}{(1+i)^k} - \sum_{k=1}^{k^*} \frac{P_k}{(1+i)^k} = \sum_{k=1}^{k^*} \left[\frac{B_{k-1}}{(1+i)^{k-1}} - \frac{B_k}{(1+i)^k} \right]$$

$$\sum_{k=1}^{k^*} \frac{R_k}{(1+i)^k} - \sum_{k=1}^{k^*} \frac{P_k}{(1+i)^k} = \left[\frac{B_0}{(1+i)^0} - \frac{B_1}{(1+i)^1} \right] + \cdots + \left[\frac{B_{k-1}}{(1+i)^{k-1}} - \frac{B_{k^*}}{(1+i)^k} \right]$$

$$= \frac{B_0}{(1+i)^0} - \frac{B_{k^*}}{(1+i)^{k^*}}.$$

Multiplying this expression by $-(1+i)^{k^*}$ and noting that $B_0 = P_0 - R_0$ provides the equation for B_{k^*}:

$$\boxed{B_{k^*} = \sum_{k=0}^{k^*} (P_k - R_k)(1+i)^{k^*-k},} \tag{3-2}$$

where $P_k - R_k$ is the **net principle** at the end of the kth interest period. This equation says that *at* TP_{k^*} *the borrower owes the lender the future worth of the net principles in the range 0 to* k^*.

Computation Note: *When the net principles are irregular, it's simpler and more informative to calculate an intermediate balance recursively. See Subsection 3.1.3.*

When a loan is a **1*P*-*A*** loan, Equation (3-2) reduces to

$$\boxed{B_{k^*} = P(F/P, i, k^*) - A(F/A, i, k^*).} \tag{3-3}$$

Present Worth Form of the IBE

To obtain a present worth equation for B_{k^*}, we sum Equation (3-1) from k^*+1 to N:

$$\sum_{k=k^*+1}^{N} \frac{R_k - P_k}{(1+i)^k} = \sum_{k=k^*+1}^{N} \left[\frac{B_{k-1}}{(1+i)^{k-1}} - \frac{B_k}{(1+i)^k} \right]$$

$$= \left[\frac{B_{k^*}}{(1+i)^{k^*}} - \frac{B_{k^*+1}}{(1+i)^{k^*+1}} \right] + \cdots + \left[\frac{B_{N-1}}{(1+i)^{N-1}} - \frac{\overbrace{B_N}^{B_N \equiv 0}}{(1+i)^N} \right]$$

$$= \frac{B_{k^*}}{(1+i)^{k^*}}.$$

Multiplying this equation by $(1+i)^{k^*}$ yields

$$\boxed{B_{k^*} = \sum_{k=k^*+1}^{N} \frac{R_k - P_k}{(1+i)^{k-k^*}},} \tag{3-4}$$

where $R_k - P_k$ is the **net payment** at the end of the kth interest period. This equation says that *at TP_{k^*} the borrower owes the lender the present worth of the remaining series of future net payments.*

Computation Note: *When the net payments are irregular, it is simpler and more informative to calculate an intermediate balance recursively.*

When a loan is a **1*P*-*A*** loan, Equation (3-4) reduces to

$$\boxed{B_{k^*} = A(P/A, i, [N-k^*]).} \tag{3-5}$$

EXAMPLE 3.1 Calculating an Intermediate Balance of a Loan (*B*-Class)

Problem: *Calculate in two ways the balance of the loan in Example 2.11 at the end of the 12th month.* NOTE: $N = 48$.
Solution: To calculate the intermediate balance B_{12} of this 1*P*-*A* loan, we can use Equation (3-3) or Equation (3-5).

Using Equation (3-3)

Substituting the values $i\% = 0.5\%$, $P = \$24{,}000$, and $A = \$563.64$ from Example 2.11 into Equation (3-3) yields the value of B_{12}:

$$B_{12} = \$24{,}000\underbrace{(F/P, 0.5\%, 12)}_{1.06168} - \$563.64\underbrace{(F/A, 0.5\%, 12)}_{12.33556} = \underline{\underline{\$18{,}527.51.}}$$

Using Equation (3-5)

Substituting the value $N = 48$ and the values of $i\%$ and A into Equation (3-5) yields the value of B_{12}:

$$B_{12} = \$563.64\underbrace{(P/A, 0.5\%, [48-12])}_{32.87102} = \underline{\underline{\$18{,}527.42.}}$$

Except for the negligible factor round-off errors, this result is identical to that obtained with Equation (3-3).

Computation Note: *When a loan is a 1P-A loan, Equation (3-5) requires less computation than Equation (3-3).*

3.1.2 Annual Interests of a Home Mortgage

Homebuyers usually purchase a house with a long-term loan called a **mortgage**. Most mortgages are 1*P*-*A* loans with terms of 15, 20, or even 30 years and a *constant* monthly interest rate. The rest of this subsection explains how to calculate the annual interest paid in each year of the term T (measured in years) of a mortgage.

Annual Interest Equation

The *annual* interest I_y paid in the yth year of a mortgage is given by

$$I_y = 12A - \Delta B_y \qquad y = 1, 2, \ldots, T, \tag{3-6}$$

where A is the monthly mortgage payment and ΔB_y is the change in the *year-end* balances

$$\Delta B_y \equiv B_{y-1} - B_y. \tag{3-7}$$

Year-End Balance Equation

Since the mortgage is a 1*P*-*A* loan, we can calculate its balance at the end of the yth year with Equation (3-5):

$$B_y = A\left(P/A, i, 12[T - y]\right) \qquad y = 1, 2, \ldots, T, \tag{3-8}$$

where the number of monthly interest periods remaining *after* the end of the yth year is $12[T - y]$.

It's easy to calculate Equations (3-8), (3-7), and (3-6) sequentially in a spreadsheet to obtain the *year-end balances* B_y and *annual interest payments* I_y of a mortgage. The ***MORTGAGE BALANCES & INTERESTS Calculators***, Excel spreadsheets in the *Book Templates* file, do this automatically.

EXAMPLE 3.2 Annual Interests of a Home Mortgage (*B*-Class)

Problem: A couple purchased a home with a 10% down payment and a $150,000 mortgage that has an *APR* of 7% compounded monthly. The term of the loan is 30 years. *Calculate the annual interests and total interest of the mortgage.*

Solution: It follows from the definition of the *APR* that

$$i\% = \frac{APR}{m} = \frac{7}{12} = 0.58333\% \text{ per month.}$$

We use the ***30-YEAR MORTGAGE BALANCES & INTERESTS Calculator*** (in the *Book Templates* file) displayed in Spreadsheet 3.1. It determines the *annual* interests of a 30-year mortgage for *arbitrary* values of P and the *APR*. The value of $i\%$ is calculated in cell E5. The label cells in array C7:E7 show that B_y, ΔB_y, and I_y are calculated with Equations (3-8), (3-7), and (3-6) in columns C, D, and E, respectively. Formula cell E39 shows that the total interest is $209,263.35.

Column F lists each **percentage normalized annual interest**—the percentage of the annual payment ($12A$) that pays the annual interest I_y. *During the first 15 years it ranges from 87.28% to 66.19%; this results in large income tax deductions.* Near the end of the term, most of each yearly payment reduces the loan balance. Figure 3.1 is a plot of the data in column F over the 30-year term.

	A	B	C	D	E	F
1						
2		***30-YEAR MORTGAGE BALANCES & INTERESTS Calculator***				
3				***P***	***APR***	***T*, a**
4			***Input:***	$150,000	7.00%	30
5			***Data:***	***A***	0.58333%	*i%* = E4/12
6				$997.95	=D4*A_P(E5,12*F4)	
7		***y*, a**	$B_y = A(P/A, i\%, 12[T - y])$	$\Delta B_y = B_{y-1} - B_y$	$I_y = 12A - \Delta B_y$	$100 \frac{I_y}{12A}$
8		0	150,000.00			
9		1	148,476.29	1,523.71	10,451.73	87.28
10		2	146,842.42	1,633.86	10,341.58	86.36
11		3	145,090.44	1,751.98	10,223.47	85.37
12		4	143,211.82	1,878.63	10,096.82	84.31
13		5	141,197.38	2,014.43	9,961.01	83.18
14		6	139,037.33	2,160.06	9,815.39	81.96
15		7	136,721.12	2,316.21	9,659.24	80.66
16		8	134,237.48	2,483.65	9,491.80	79.26
17		9	131,574.29	2,663.19	9,312.26	77.76
18		10	128,718.58	2,855.71	9,119.73	76.15
19		11	125,656.42	3,062.15	8,913.29	74.43
20		12	122,372.91	3,283.51	8,691.93	72.58
21		13	118,852.03	3,520.88	8,454.57	70.60
22		14	115,076.63	3,775.40	8,200.04	68.47
23		15	111,028.30	4,048.33	7,927.12	66.19
24		16	106,687.32	4,340.98	7,634.46	63.75
25		17	102,032.53	4,654.79	7,320.65	61.13
26		18	97,041.24	4,991.29	6,984.16	58.32
27		19	91,689.13	5,352.11	6,623.34	55.31
28		20	85,950.12	5,739.01	6,236.43	52.08
29		21	79,796.23	6,153.89	5,821.56	48.61
30		22	73,197.48	6,598.75	5,376.69	44.90
31		23	66,121.71	7,075.78	4,899.67	40.91
32		24	58,534.42	7,587.28	4,388.16	36.64
33		25	50,398.65	8,135.77	3,839.68	32.06
34		26	41,674.75	8,723.90	3,251.54	27.15
35		27	32,320.19	9,354.56	2,620.89	21.89
36		28	22,289.40	10,030.80	1,944.65	16.24
37		29	11,533.47	10,755.92	1,219.52	10.18
38		30	**0.00**	11,533.47	441.97	3.69
39				**=Sum(E9:E38)**	**$209,263.35**	

Spreadsheet 3.1 ***30-YEAR MORTGAGE BALANCES & INTERESTS Calculator*** NOTES: (1) Label cell E6 explains how *A* is calculated. (2) The columns of formula values are easily obtained by copying the formula in row 8 or 9 of the column and pasting it down the column to the cell in row 38. (3) The label cells in array C7:F7 explain what is calculated in their columns. (4) *As required,* $B_{30} = 0$. (5) ***The total interest is 139.5% of P.***

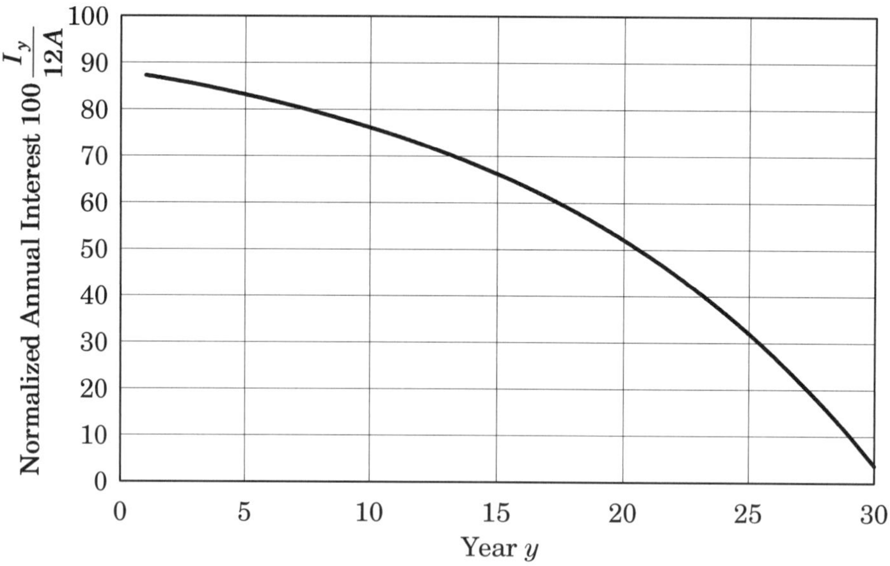

Figure 3.1 **Yearly Variation of the Normalized Annual Interest of a 30 Year Mortgage** NOTE: The *APR* is 7% compounded monthly.

3.1.3 Recursive Calculation of Intermediate Balances

Motivation

We can use Equation (3-2) or (3-4) to *calculate one loan*[1] *balance,* and for a *1P-A* loan we can use Equation (3-3) or (3-5). Example 3.1 illustrates the calculations for a *1P-A* loan. When the cash flows of a loan are ***irregular***, calculating its balances is tedious. Hence, we use a recursive function to calculate some or all of the intermediate balances of any loan.

Recursive Balance Function

A **simple recursive function** $f(k)$ is a function—of a nonnegative integer argument k—whose value depends on (1) its *previous* value $f(k-1)$, (2) a *specified* value of $f(0)$, and (3) an optional *specified* function of k.

The Balance Equation (2-31) is a simple recursive function for the balance of *any* loan or simple investment:

$$\boxed{B_k = B_{k-1}(1+i) + P_k - R_k \qquad k = 1, 2, \ldots, N,} \tag{3-9}$$

Substituting the definition of a cash flow from Equation (2-108),

$$CF_k \equiv R_k - P_k, \tag{3-10}$$

into Equation (3-9) yields the **Recursive Balance Function** (RBF):

$$\boxed{B_k = B_{k-1}(1+i) - CF_k \qquad k = 1, 2, \ldots, N.} \tag{3-11}$$

[1] In this book "loan" and "simple investment" are synonymous.

For example, when B_0 and the first four values of CF_k are *specified*, we obtain B_4 by performing the following calculations *sequentially*:

$$\begin{aligned} B_1 &= B_0(1+i) - CF_1, \\ B_2 &= B_1(1+i) - CF_2, \\ B_3 &= B_2(1+i) - CF_3, \\ B_4 &= B_3(1+i) - CF_4. \end{aligned}$$

Computation Note: *We use Excel to conveniently calculate intermediate balances with the RBF.*

EXAMPLE 3.3 Recursive Calculation of Intermediate Loan Balances (*B*-Class)

Problem: *Recursively calculate the balance of the loan in Example 3.1 at the end of the 12th month.* NOTE: $B_0 = \$24{,}000$, and $A = \$563.64$.

Solution: Spreadsheet 3.2 uses the RBF to calculate B_{12}. Except for the negligible factor round-off errors, this result is identical to those in Example 3.1.

	A	B	C	D	E
1					
2		***RECURSIVE CALCULATION of 12 LOAN BALANCES***			
3				***i%***	$\boldsymbol{B_0}$
4			***Input:***	0.5%	$24,000
5				***Solution:***	
6		***k***	$\boldsymbol{CF_k}$**,$**	$\boldsymbol{B_k = B_{k-1}(1+i) - CF_k}$	
7		0	(24,000)	$24,000	
8		1	563.64	23,556.36	= D7*(1+D$4)-C8
9		2	563.64	23,110.50	
10		3	563.64	22,662.41	
11		4	563.64	22,212.09	
12		5	563.64	21,759.51	
13		6	563.64	21,304.66	Copy Cell D8 and Paste Down Array D9:D19
14		7	563.64	20,847.55	
15		8	563.64	20,388.15	
16		9	563.64	19,926.45	
17		10	563.64	19,462.44	
18		11	563.64	18,996.11	
19		12	563.64	**$18,527.45**	

Spreadsheet 3.2 NOTES: (1) This is a recursive calculation of the first 12 intermediate balances of the loan in Example 3.1. (2) The balance formulas in array D9:D19 are obtained by copying the balance formula cell D8 and pasting it down array D9:D18.

3.2 OTHER INTEREST RATES

3.2.1 Annual Percentage Yield

Effective Annual Interest Rate

Two or more interest rates or APRs are not comparable! They *often* do not indicate which earns the most interest, and they *never* indicate by how much. For example, suppose bank X offers a one-year CD with an *APR* of 6% compounded yearly, and bank Y has one with an *APR* of 5.9% compounded monthly. *Which APR earns more interest?*

To compare the interest earning capacities of interest rates or *APR*s we use the following form of Equation (2-16):

$$\frac{I_{\mathrm{T}}}{P} = (1+i)^m - 1, \tag{3-12}$$

where m is the number of interest periods in a *year*. It calculates the *normalized total annual interest* of a $1P$-$1R$ loan or investment. *This ratio is the true measure of the interest-earning capacities of interest rates and APRs.* Hence, we define the **effective annual interest rate** i_{e} of an interest rate i by

$$i_{\mathrm{e}} \equiv \frac{I_{\mathrm{T}}}{P} = (1+i)^m - 1, \tag{3-13}$$

and *we compare the interest-earning capacities of interest rates and APRs by their values of* i_{e}.

We can now compare the bank *APR*s of 5.9% compounded monthly and 6% compounded annually:

$$i_{\mathrm{e},X}\% = 100\left(\frac{I_{\mathrm{T}}}{P}\right)_X = 100\left[\left(1+\frac{APR}{100m}\right)^m - 1\right]_X = 100\left[\left(1+\frac{6}{100\times 1}\right)^1 - 1\right]_X = \underline{\underline{6\%}}$$

$$i_{\mathrm{e},Y}\% = 100\left(\frac{I_{\mathrm{T}}}{P}\right)_Y = 100\left[\left(1+\frac{APR}{100m}\right)^m - 1\right]_Y = 100\left[\left(1+\frac{5.9}{100\times 12}\right)^{12} - 1\right]_Y \underline{\underline{6.0622\%}}.$$

As you can see, an *APR* of 5.9% compounded monthly earns *more* interest than 6% compounded yearly. Weekly and daily compounding earn even more.

Definition of the Annual Percentage Yield

The **annual percentage yield** *APY* is *defined* as[2]

$$\boxed{APY \equiv 100 i_{\mathrm{e}} = \begin{cases} 100\left[(1+i)^m - 1\right] \\ 100\left[\left(1+\dfrac{APR}{100m}\right)^m - 1\right]. \end{cases}} \tag{3-14a,b}$$

[2]On May 29th, 1968, the United States Congress passed the Truth in Lending Act. Prior to its enactment, lenders were not required to disclose the *APY*s of their loans.

Table 3.1 shows the variation of the *APY* with the *APR* and its compounding frequency. *The difference between the interest earned with frequent compounding and annual compounding increases with the frequency of compounding and the APR*. For example, when the *APR* is 4%, daily compounding earns about 2% more interest than annual compounding. However, when the *APR* is 20%, daily compounding earns over 10% more interest than annual compounding.

TABLE 3.1 Effect of the *APR* and Its Compounding Frequency on the *APY*

COMPOUNDING FREQUENCY *m*	*APY*					
	APR = 4%	*APR* = 8%	*APR* = 12%	*APR* = 16%	*APR* = 20%	*APR* = 50%
Annually 1	4	8	12	16	20	50
Semiannually 2	4.04	8.16	12.36	16.64	21.	56.25
Quarterly 4	4.060	8.243	12.551	16.986	21.551	60.181
Monthly 12	4.074	8.300	12.683	17.227	21.939	63.209
Weekly 52	4.079	8.322	12.734	17.322	22.093	64.479
Daily 365	4.081	8.328	12.747	17.347	22.134	64.816

EXAMPLE 3.4 Calculating the *APY* of a Spec House (*SI*-Class)

Problem: A **spec house** is built on a speculative basis without a buyer. The monthly cash flows of a spec house are listed in Table 3.2. NOTE: A buyer purchased the house at the end of the 14th month. *Calculate the APY of this investment. (2) Calculate the balances of the investment.*

TABLE 3.2 **Monthly Cash Flows of a Spec House**[1]

MONTH	TRANSACTION	CASH FLOW, $
0	Land	–125,000
1	House Plans	–2,500
2	Survey	–1,000
3	Building Permit[1]	–150
4	Foundation	–20,000
5	Framing, Roof, Windows & Siding	–75,000
6	Rough Plumbing & Electrical	–21,000
7	Sheetrock, Tape & Spackle	–16,000
8	Flooring	–22,000
9	Plumbing & Electrical Fixtures	–10,000
10	Kitchen Cabinets, Fixtures & Appliances	–30,000
11	Landscaping & Decking	–3,800
14	Payment from Sale of the House	405,000

[1]Construction can't start until a building permit is issued. This may take several months; it depends on the locality and its economy, which affects the demand for building permits.

Solution: Since there is only one change in the signs of the cash flows, the spec house is a simple investment. Spreadsheet 3.3 shows that the *monthly* $i^*\%$ equals 2.1499%. In addition, all of the balances except the last one, B_{14}, are positive. The last balance of *any* simple investment or loan is always zero.

We use Equation (3-14a) to convert $i^*\%$ to the *APY*:

$$APY = 100[(1+i^*)^m - 1] = 100[(1+0.021499)^{12} - 1] = \underline{\underline{29.08\% \text{ per year.}}}$$

	A	B	C	D	E
1					
2		***MONTHLY i*% & BALANCES of the SPEC HOUSE INVESTMENT***			
3					***Solution:***
4		***k***		**CF_k,$**	$B_k = B_{k-1}(1+i^*) - CF_k$
5		0		(125,000)	125,000
6		1		(2,500)	130,187.41
7		2		(1,000)	133,986.35
8		3		(150)	137,016.96
9		4		(20,000)	159,962.73
10		5		(75,000)	238,401.82
11		6		(21,000)	264,527.29
12		7		(16,000)	286,214.45
13		8		(22,000)	314,367.86
14		9		(10,000)	331,126.55
15		10		(30,000)	368,245.53
16		11		(3,800)	379,962.56
17		12		0	388,131.49
18		13		0	396,476.04
19		14		405,000	**0.00**
20			**=IRR(D5:D19)**	**2.1499%**	

Spreadsheet 3.3 NOTES: (1) The end-of-month balances are recursively calculated with the RBF, Equation (3-11), in formula array E6:E19. (2) Like the balances of *any* simple investment, all of the balances except the last one are *positive*, and it *always* equals 0.

3.2.2 Simple Interest

A loan earns **simple interest** when interest is *not* earned on unpaid interest. For example, consider a $1P$-$1R$ loan with simple interest. The interest earned, but not paid, in each interest period is just Pi. At the end of the term, the payment F is the sum of the N earned interests and the principle:

$$F = P + NPi. \tag{3-15}$$

Subtracting P from both sides of this equation yields the **total interest** earned by a $1P$-$1R$ loan **with simple interest** I_{T_S} :

$$I_{T_S} = NPi. \tag{3-16}$$

When a $1P$-$1R$ loan earns compound interest, the total interest I_{T_C} is given by

$$I_{T_C} = P[(1+i)^N - 1]. \tag{3-17}$$

Subtracting Equation (3-16) from Equation (3-17) and dividing the result by P yields the **normalized extra interest** $\Delta I_T P^{-1}$ earned with compound interest:

$$\frac{\Delta I_T}{P} \equiv \frac{I_{T_C} - I_{T_S}}{P} = (1+i)^N - (1+Ni). \tag{3-18}$$

Figure 3.2 is a plot of $\Delta I_T P^{-1}$ versus N with $i\%$ as a parameter. When $i\% = 9\%$ per period, the *extra* interest earned with compound interest is about 50% of P after 10 interest periods. This dramatically illustrates ***the power of compound interest.***[3]

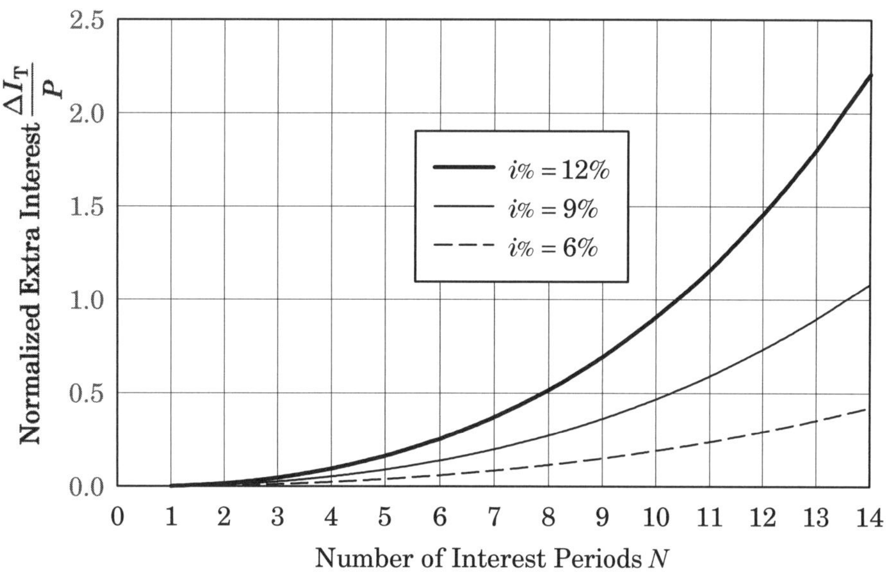

Figure 3.2 **Effects of N and $i\%$ on the Normalized Extra Interest of $1P$-$1R$ Loans**

3.2.3 Continuous Compounding

Effective Annual Interest Rate

Occasionally a lending institution will offer an *APR* that is **compounded continuously**. This means that the interest period is infinitesimally small. The

[3]The following quotation has been attributed to Albert Einstein: ***"Compound interest is the most powerful force in the universe."***

effective annual interest rate i_∞ for **continuous compounding** is *defined* as the limit of the effective annual interest rate i_e, defined in Equation (3-13), as m approaches infinity:

$$i_\infty \equiv \lim_{m\to\infty} i_e = \lim_{m\to\infty} (1+i)^m - 1. \qquad (3\text{-}19)$$

To evaluate this limit, we first substitute the expression for i from Equation (2-6),

$$i = \frac{r}{m},$$

where r is the **nominal interest rate,** $\frac{APR}{100}$, into Equation (3-19):

$$i_\infty \equiv \lim_{m\to\infty} i_e = \lim_{m\to\infty}\left(1+\frac{r}{m}\right)^m - 1.$$

Multiplying the exponent in this equation by rr^{-1} yields the limit expression for the mathematical constant e raised to rth power:

$$i_\infty = \lim_{m\to\infty}\left(1+\frac{r}{m}\right)^{\frac{mr}{r}} - 1 = \left[\lim_{m\to\infty}\left(1+\frac{r}{m}\right)^{\frac{m}{r}}\right]^r - 1.$$

Hence,

$$\boxed{i_\infty = e^r - 1.} \qquad (3\text{-}20)$$

When interest is compounded continuously, the APY is *defined* as

$$APY_\infty \equiv 100 i_\infty = 100(e^r - 1) = 100(e^{APR_\infty/100} - 1). \qquad (3\text{-}21)$$

Table 3.3 shows the variation of the APY with the APR when interest is compounded daily and continuously. As you can see, ***the difference between daily and continuous compounding is negligible over a wide range of APRs***.

TABLE 3.3 Comparison of the *APY*s for Daily and Continuous Compounding

COMPOUNDING FREQUENCY *m*	*APY*					
	APR = 4%	*APR* = 8%	*APR* = 12%	*APR* = 16%	*APR* = 20%	*APR* = 50%
Daily: 365	4.081	8.328	12.747	17.347	22.134	64.816
Continuously: ∞	4.081	8.329	12.750	17.351	22.140	64.872

Continuous Cash Flow Model

It should be clear from Table 3.3 that using continuous compounding in calculations yields results that are essentially identical to those when interest is compounded daily. That being said, what is the practical use of continuous compounding? ***It is***

used to model cash flow series that can't be conveniently represented by a combination of the three basic series. The series are represented by *continuous* functions on the timeline and are called **cash flow densities**. Their use with continuous compounding is called the **continuous cash flow model**.

This model has two advantages over the *simple* discrete cash flow model that we have been using: *(1) We can use a wide variety cash flow densities f(t) to represent series. (2) The End-of-Year Convention is unnecessary because series are, by definition, continuously distributed on the timeline.*

To calculate *APY* for an investment with the continuous cash flow model, we equate the present worths of the disbursement and receipt cash flow densities and solve for the *unknown* nominal interest rate *r*. The equality is a cumbersome *transcendental* (non-algebraic) equation that can be solved with mathematics software, such as *Mathematica*. Excel's Solver command can be used, but a cumbersome one-dimensional transcendental equation is extremely *tedious* to input.

Practice Note:*The continuous cash flow model is occasionally used in the chemical and financial industries. However, the modest increase in the accuracy of the APY of a proposed investment will not normally change its funding decision. Hence, simple is better, and the continuous cash flow model is not used in this book.*

3.3 COMPOSITE SERIES

3.3.1 Introduction

A **composite series** is the sum of the cash flows of two or more basic series, which are its **components**. The basic series may be shifted, nonshifted, or both. For example, Figure 3.3 shows the cash flows of an $A + G + A_{1,S}$ composite series. Its components are a uniform series, a gradient series, and a *decreasing* geometric gradient series that is *shifted* N interest periods into the future. The subscript "S" denotes the shift. The SPT of $A_{1,S}$ is TP_N.

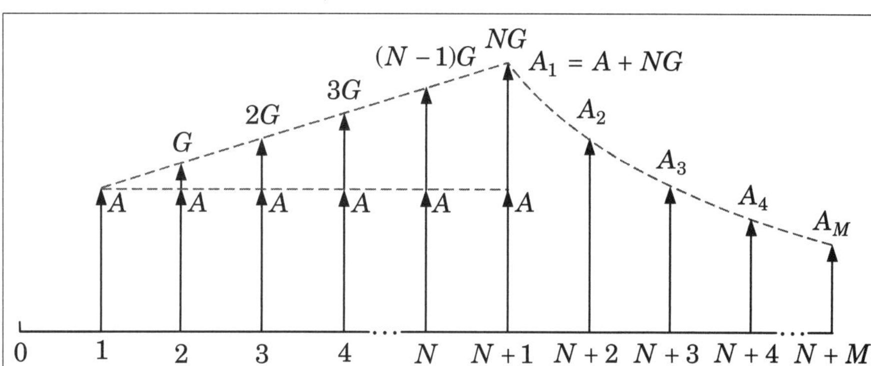

Figure 3.3 **A Composite Series with Three Components** NOTE: A_1 is the *last* cash flow of the sum of the uniform and gradient series; that is, $A_1 = A + NG$.

The **present** (or future) **worth** of a **composite series** is the sum of the present (or future) worths of its components at the PT (or FT). For example, the present worth of the $A + G + A_{1,S}$ series is given by

$$\boxed{PW_{A+G+A_1,S} = A(P/A,i,N) + G(P/G,i,N) + A_1(P/A_1,g,i,M)(P/F,i,N).} \quad (3\text{-}22)$$

EXAMPLE 3.5 Making a Counteroffer for an Investment (*SI*-Class)

Problem: A small machine shop is on the market for $80,000. A buyer's aftertax cash flow at the end of the first year is estimated to be $10,000. It is assumed that this will increase by $2,000 per year to $20,000 at the end of the sixth year. However, new competition will eventually cause the aftertax cash flows to decrease by 5% per year over the remaining four years of the assumed 10-year investment life. The market value is assumed to be $40,000 at the end of the 10th year.[4] (1) *What is the IRR% of this investment?* (2) *If the IRR% is less than 18% per year, what counteroffer should a buyer make so that the IRR% is 18% per year?*

Solution: The $A + G + A_{1,S}$ series in Figure 3.3 represents the aftertax cash flows when $A = \$10{,}000$, $G = \$2{,}000$, $M = N = 5$, $A_1 = A + 5G = \$20{,}000$, and $g = -0.05$.

(1) Calculation of the IRR%

Spreadsheet 3.4 calculates an IRR% of 16.43% per year. Hence, we must calculate the counteroffer that will yield 18% per year.

	A	B	C	D	E	F	G	H
1								
2		***IRR% of a MACHINE SHOP INVESTMENT***						
3				***P***	***MV***	***A***	***G***	***g%***
4			***Input:***	$80,000	$40,000	$10,000	$2,000	-5%
5		***k***				***CF*, $**	$B_k = B_{k-1}(1 + i^*) - CF$	
6		0			=-D4	(80,000)	80,000	
7		1			=F4	10,000	83,146.93	
8		2		=F$4+(B8-1)*G$4		12,000	84,811.03	
9		3	Copy F8 and Paste Down Array F9:F12			14,000	84,748.59	
10		4				16,000	82,675.89	
11		5				18,000	78,262.57	
12		6				20,000	71,123.99	
13		7	=F$12*(1+H$4)^(B13-B$12)			19,000	63,812.27	
14		8	Copy F13 and Paste Down Array F14:F15			18,050	56,248.96	
15		9				17,148	48,345.23	
16		10	=F$12*(1+H$4)^(B16-B$12) +E4			56,290	**0.00**	
17					**=IRR(F6:F16)**	**16.43%**		

Spreadsheet 3.4 NOTE: The *Formula Notes* in the label cells explain how the cash flows in array F6:F16 are obtained. A reader will not understand the spreadsheet without them.

Computation Note: *Unlike previous calculations of the IRR%, the values of k in array B8:B16 are used to calculate the cash flows in array F8:F16.*

(2) Calculation of the Counteroffer

[4]The *estimated* market value is significantly below the asking price because the buyer is concerned about the effect of competition on the value of the investment.

A Goal Seek iteration (not shown) reveals that a counteroffer of $74,532 will yield an IRR% of 18% per year. A 16.43% IRR% is an excellent rate of return for this type of investment. Hence, the potential buyer should make a counteroffer of $74,500 to get a better IRR%. The seller may go for it or may make another counteroffer. The buyer will never know unless he/she tries the $74,500 counteroffer!

3.3.2 Decreasing Linear Series

Definitions

A **decreasing linear series** $A - G$ is a composite series that is the *difference* of a uniform series and a *decreasing* gradient series with the *same* series present and future times. Figure 3.4 shows a decreasing linear series of N cash flows.

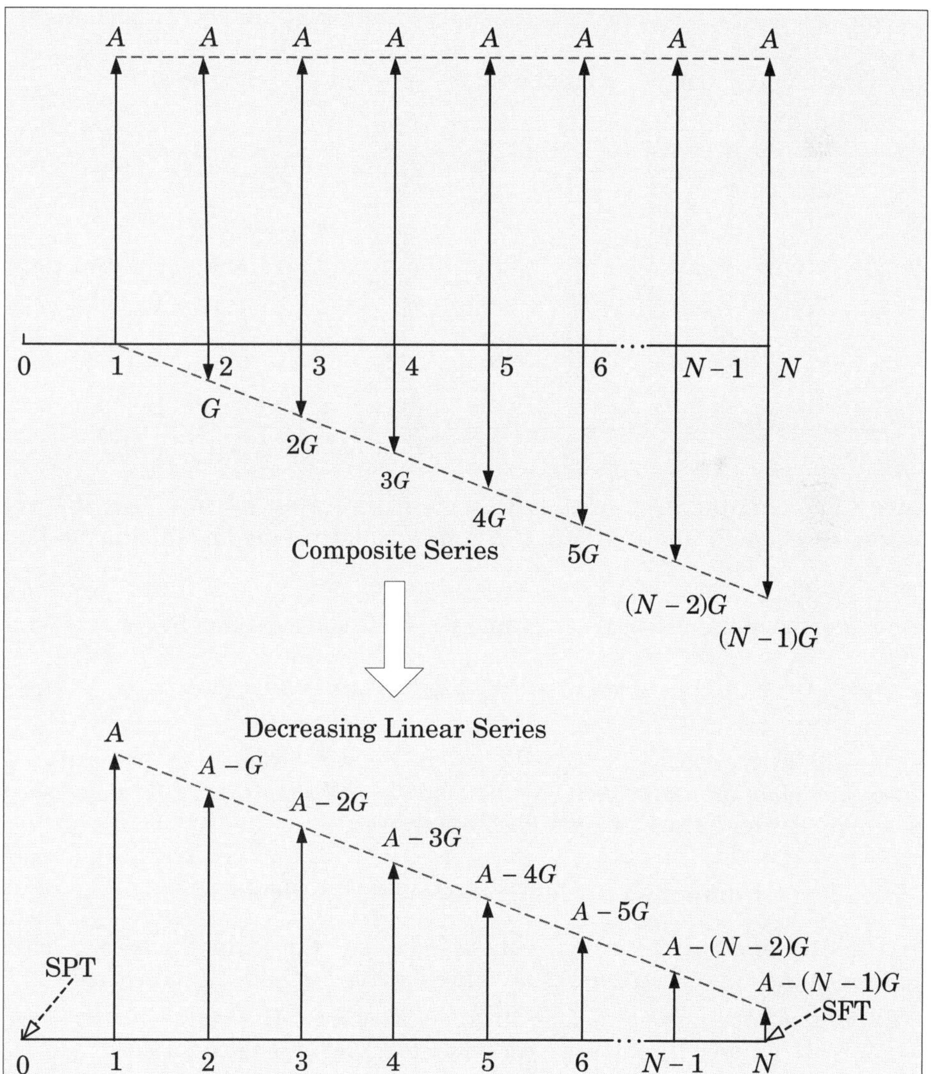

Figure 3.4 **A Decreasing Linear Series of N Cash Flows** The SPT and SFT are TP_0 and TP_N, respectively. As with the three basic series, $N = TP_N - TP_0$.

The **metrics** of a decreasing linear series are A and G. A is the magnitude of its **apex** (largest) cash flow, and G is the magnitude of each cash flow **decrement**.

Present Worth

The **present worth** of a **decreasing linear series** $PW_{A\text{-}G}$ at its SPT is just the difference of the present worths of its uniform and gradient series at the SPT:

$$\boxed{PW_{A-G} = A(P/A,i,N) - G(P/G,i,N),} \qquad (3\text{-}23)$$

where N is the number of interest periods from the SPT to the SFT.

3.3.3 Composite Series Transformation Procedure

The composite series in Figure 3.5 is the *sum* of a uniform series of N cash flows and a decreasing linear series of M cash flows that is *shifted* N interest periods into the future. This composite series is often used to represent a receipt series.

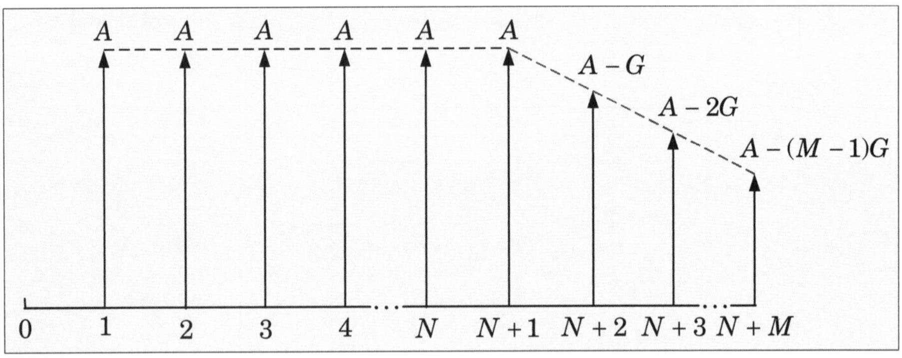

Figure 3.5 **A Composite Series with Two Components** NOTE: The SPT of the uniform series is TP_0, and the SPT of the *shifted* decreasing linear series is TP_N.

The present worth of this composite series $PW_{A+S(A-G)}$ is the sum of the present worths of the uniform series and the shifted decreasing linear series:

$$PW_{A+S(A-G)} = A(P/A,i,N) + \left[A(P/A,i,M) - G(P/G,i,M)\right](P/F,i,N). \qquad (3\text{-}24)$$

We can simplify the calculation of $PW_{A+S(A-G)}$ by transforming the composite series into an *equivalent* composite series with a simpler PW equation. To do so, we use the **Composite Series Transformation Procedure**:

Composite Series Transformation Procedure
(1) Add an appropriate series or shifted series to the composite series. (The choice of the appropriate series or shifted series comes with *experience*.)
(2) Subtract the *same* series or shifted series from the result of Step 1 to obtain an *equivalent* composite series with a *simpler* PW equation.

Let's apply the Composite Series Transformation Procedure to obtain a simpler equation for $PW_{A+S(A-G)}$:

1. *Add* a shifted gradient series of $M-1$ cash flows to the shifted decreasing linear series in Figure 3.5. This yields a uniform series of $N+M$ cash flows (see Figure 3.6).
2. *Subtract* the shifted gradient series from this uniform series to obtain an equivalent composite series with a simpler *PW* equation (see Figure 3.6):

$$\boxed{PW_{A+S(A-G)} = A(P/A,i,N+M) - G(P/G,i,M)(P/F,i,N).} \tag{3-25}$$

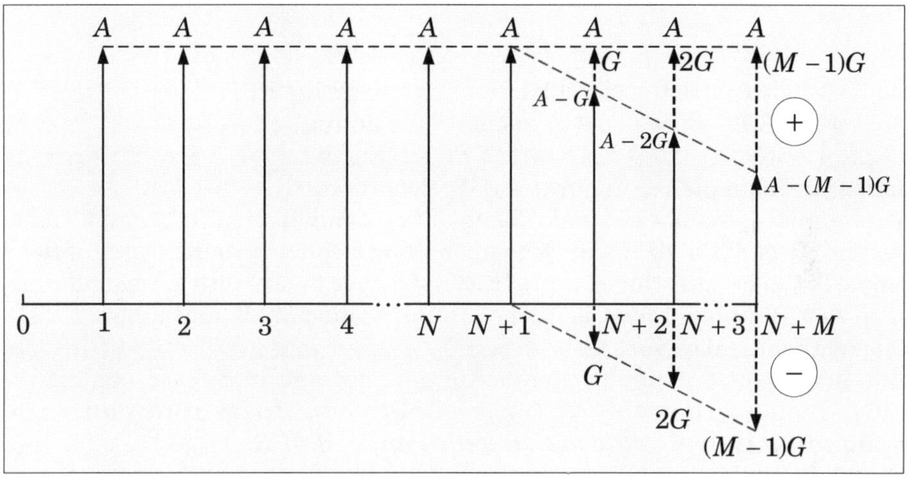

Figure 3.6 **Transformation of the Composite Series in Figure 3.5** NOTES: (1) The equivalent composite series has a simpler *PW* equation. (2) Table 3.4 compares the number of factors and arithmetic operations required to calculate the present worth with Equations (3-24) and (3-25).

Computation Note: *Equation (3-24) requires four factors and three worth operations, and Equation (3-25) requires three factors and two worth operations:*

TABLE 3.4 Worth Comparisons of the Sum of a Uniform Series and a Decreasing Linear Series

FACTORS and OPERATIONS	Eq. (3-24)	Eq. (3-25)
Number of Factors	4	3
Worth Differences	1	1
Worth Sums	1	0
Worth Products	1	1

Remark: If you understand how to transform the sum of a uniform series and a shifted decreasing linear series into an equivalent composite series with a simpler present worth equation, you should have no difficulty in transforming other composite series.

3.4 SUMMARY

Subsection 3.1.1 derives the future and present worth forms of the Intermediate Balance Equation (IBE) for a loan (or simple investment). When the loan is a 1*P*-1*A* loan, the future worth and present worth forms reduce to Equations (3-3) and (3-5), respectively. Example 3.1 illustrates their use. However, when the cash flows are irregular, we use the **Recursive Balance Function** (RBF) derived in Subsection 3.1.3:

$$\boxed{B_k = B_{k-1}(1+i) - CF_k \qquad k = 1, 2, \ldots, N.} \tag{3-11}$$

Example 3.3 illustrates the use of Excel to implement the RBF.

Subsection 3.1.2 shows how to calculate the annual interests of a 30-year home mortgage with the ***30-YEAR MORTGAGE BALANCES & INTERESTS Calculator***. Example 3.2 illustrates the calculation.

The annual percentage yield *APY* is defined Subsection 3.2.1, and Table 3.1 shows the effect of the *APR* and its compounding frequency on the value of the *APY*. Example 3.4 calculates the *monthly* IRR% of a spec house with Excel and converts it to an *APY*. Simple interest is defined in Subsection 3.2.2, and Figure 3.2 shows the normalized extra interest earned by using compound interest instead of simple interest. Continuous compounding is defined in Subsection 3.2.3, and Table 3.3 shows that ***the difference between daily and continuous compounding is negligible over a wide range of APRs***.

Composite series are defined Section 3.3, and Example 3.5 illustrates the use of the composite series displayed in Figure 3.3. The decreasing linear series is defined in Subsection 3.3.2 and illustrated in Figure 3.4. Equation (3-23) calculates its present worth. The Composite Series Transformation Procedure is defined in Subsection 3.3.3, and Figure 3.6 and Equation (3-25) illustrate its use.

3.5 PROBLEMS

(P3-1) (Part A) *Calculate the balance of the following investment at the end of the fifth interest period in three ways.* **(Part B)** *Calculate all of the balances.*

CASH FLOW, $\$10^3$	−8	2	2	2	2	2	2	2	2	2
YEAR *k*	0	1	2	3	4	5	6	7	8	9

(P3-2) A mortgage has a term of 15 years and an *APR* of 4.5% compounded monthly. *Calculate the total interest paid over the term.*

(P3-3) *Calculate the APY and balances of the spec house in Example 3.4 when the Rough Plumbing & Electrical cost is $29,500.*

(P3-4) *Which of the following APRs has the largest APY: 4% compounded monthly, 3.9% compounded weekly, 3.85% compounded continuously.* Show the calculation of each *APY*.

(P3-5) A repayment plan of a loan consists of a uniform series of 10 \$400 cash flows followed by a decreasing linear series of five cash flows with the first one being the metric of the uniform series. The interest rate is 6% per year. *What is the present worth of the series if the decrement of the decreasing linear series is \$50?* Calculate the present worth in two ways.

4

Annual Worth Comparisons

OBJECTIVES

Engineers prepare the equipment specifications in requests for quotations from vendors, and they perform the technical and *economic* evaluations of the vendor bids. After reading this chapter you will be able to perform an *annual worth comparison of these* ***cost alternatives***—an ***AWC*-Class** problem—to

1. Find the least-cost alternative of two or more cost alternatives with the same or different lives, component replacements, and a specified service time.
2. Find the least-cost alternative when the lives of *two* cost alternatives are different and the service time is the least common multiple of their lives.
3. Find the least-cost alternative for three other classes of cost alternatives: Extra Capacity, Dissimilar, and Irregular.
4. Find the least-cost alternative when two or more cost alternatives provide perpetual service.
5. Make annual worth investment equivalence calculations.

In Chapter 5 we use present worth comparisons and calculations to accomplish the same tasks; we also compare the two worth comparison methods.

4.1 INTRODUCTION

4.1.1 Definitions

Cost alternatives satisfy the *same* need, such as compressing a gas, heating a fluid or printing engineering drawings. They have cash flows that are predominately *disbursements. Choosing the least-cost alternative precludes selecting additional ones.* Hence, we say that cost alternatives are **mutually exclusive**. A **least-cost study** identifies the **least-cost alternative**—the one with the *smallest* worth of its cash flows. The **cash flow elements** of a cost alternative are its series, such as a uniform and gradient series, and its single cash flows, such as the purchase price and salvage value.

Computation Note: *We define disbursements as positive cash flows in least-cost calculations to avoid cluttering them with negative numbers.*

The interest rate used in a least-cost study is called the **minimum attractive rate of return** (MARR) r^*. It is the minimum IRR that a *proposed* investment must earn to be considered for funding. The MARR depends on the type and magnitude of the investment, its risk, and the source(s) of investment capital. *The value used by an engineer in a least-cost study is normally specified by a client or a financial officer in his/her company.*

Book Convention: *Since the timespan used in a least-cost study is normally measured in* ***years****, the MARR% is specified as an* ***annual*** *interest rate. And we follow conventional practice and replace the terminology "uniform worth" used in Chapter 2 with "****annual worth****" in least-cost studies.*

The **life** L of a cost alternative is the timespan on which it meets its performance specifications, and the **service time** ST is the timespan on which it is required. The service time is specified by the engineer tasked with a least-cost study or his/her manager. A **perpetual service time** denotes an *ongoing* requirement for an alternative. For example, the Golden Gate Bridge and the Lincoln Tunnel have perpetual service times.

4.1.2 Rationale & Consistency of Annual Worth Comparisons

Rationale

The following is the annual worth form of the Uniform Worth Comparison Rules stated in Subsection 2.7.1:

Annual Worth Comparison Rule
AW-Rule: To compare the worths of two or more alternatives, their annual worths must be calculated on the common annual service time:ST.

In Examples 2.16 and 2.18 we compared the *present* worths of the costs and salvage values of compressors X and Y, and in Example 2.9 we compared the present worths of two sweepstakes prizes. *However, present or future worth calculations may be difficult to understand without a background in engineering economics.* On the other hand, most adults can prepare a budget to monitor monthly incomes and expenses. Therefore, ***the AW-Rule is a natural way to compare the cash flows of cost alternatives.***[1] To apply the *AW-Rule*, we **annualize** the cash flows of the alternatives; that is, *we convert them to equivalent annual series on the annual interest periods in the service time.*

Consistency Equation

Annual worth and present worth least-cost comparisons are **consistent**; that is, both comparisons identify the same least-cost alternative. This follows from the **Consistency Equation**:

$$\boxed{AW_X = PW_X(A/P, r^*, ST),} \tag{4-1}$$

where X is *any* alternative and $(A/P, r^*, ST)$ is the *common* capital recovery factor.

[1] In the remainder of this chapter "alternative" means "cost alternative."

Net Annual Worth of a Geometric Gradient Series

When a cash flow element is a geometric gradient series of costs, the first cost A_1 of the series is often covered by a manufacturer's warranty. Hence, we must subtract the annual worth of A_1 from the annual worth of the series. We call the result the **net annual worth** NAW_{A_1} of a **geometric gradient series**:

$$\boxed{NAW_{A_1} \equiv A_1\left[\left(A/A_1, g, i, ST\right) - \left(P/F, i, 1\right)\left(A/P, i, ST\right)\right].} \tag{4-2}$$

4.2 SPECIFIED SERVICE TIME

4.2.1 Introduction

Example 4.1 illustrates the use of Equations (4-1) and (4-2) with a specified service time.

EXAMPLE 4.1 Finding the Least-Cost Compressor: An Annual Worth Comparison When $ST = L = 10a$ (*AWC*-Class)

Problem: Table 4.1 lists the economic data that was used for compressors X and Y in Example 2.18. The annual maintenance costs are represented by a geometric gradient series, and the MARR% is 20% per year. *(1) Use an annual worth comparison to find the least-cost compressor when the maintenance cost of each compressor in the first year is covered by its manufacturer's warranty. (2) Show that the results are consistent with those of Example 2.18.*

TABLE 4.1 Economic Data for the Compressors

DATA TYPE	*X*	*Y*
Purchase Price: P	\$80,000	\$85,000
Salvage Value: S	\$4,000	\$4,500
Annual Energy Cost: A	\$3,000	\$2,200
First Year Maintenance Cost: A_1	\$750	\$460
Annual Rate of Change of Maintenance Costs: $g\%$	5%	5%
Service Time: ST	10a	

Solution: ***(1) Find the Least Cost Compressor***

Table 4.2 shows the **annual worth table** used to present the calculations. It clearly displays and contrasts the annual worths of each compressor's cash flow elements. Subsection 4.2.2 describes the ***ANNUAL WORTH COMPARISON Calculator***, an Excel spreadsheet created by the author and located in the *Book Templates* file. It *automates* annual worth comparisons of two or more cost alternatives whose lives equal or exceed the service time.

As in the present worth comparison of Example 2.18, compressor X is the least-cost alternative.

(2) Consistency Check

We now use Equation (4-1) to calculate the values of AW_X and AW_Y from the values of PW_X and PW_Y obtained in Example 2.18:

TABLE 4.2 Comparing the Costs of Compressors X and Y: $ST = L = 10$ a; $r^{*}\% = 20\%$ per Year[1] Note: *Negative numbers are enclosed in parentheses ( ).*

Worth Type	Compressor Annual Worth Table	
	X	Y
AW_P	$\$80{,}000\underbrace{(A/P, 20\%, 10)}_{0.23852} = \$19{,}082$	$\$85{,}000\underbrace{(A/P, 20\%, 10)}_{0.23852} = \$20{,}274$
AW_S	$(\$4{,}000)\underbrace{(A/F, 20\%, 10)}_{0.03852} = (\$154)$	$\$4{,}500\underbrace{(A/F, 20\%, 10)}_{0.03852} = (\$173)$
AW_A	$3,000	$2,200
NAW_{A_1}	$\$750\left[\underbrace{(A/A_1, 5\%, 20\%, 10)}_{1.17182} + -\underbrace{(P/F, 20\%, 1)}_{0.83333}\underbrace{(A/P, 20\%, 10)}_{0.23852}\right] = \$730$	$\$460\left[\underbrace{(A/A_1, 10\%, 20\%, 10)}_{1.17182} + -\underbrace{(P/F, 20\%, 1)}_{0.83333}\underbrace{(A/P, 20\%, 10)}_{0.23852}\right] = \$448$
ΣAW	**$22,658**	**$22,749**

[1] Since the first-year maintenance cost is covered by a manufacturer's warranty, $AW_{A_1} \rightarrow NAW_{A_1}$.

$$AW_X = \underbrace{PW_X}_{\$94{,}991}\underbrace{(A/P, 20\%, 10)}_{0.23852} = \underline{\underline{\$22{,}657}}$$

$$AW_Y = \underbrace{PW_Y}_{\$95{,}373}\underbrace{(A/P, 20\%, 10)}_{0.23852} = \underline{\underline{\$22{,}748}}.$$

Except for the negligible round-off errors[2] in the worth table, the consistency equation results are identical to the annual worths calculated in Table 4.2.

4.2.2 *ANNUAL WORTH COMPARISON* Calculator

Spreadsheet 4.1 shows the ***ANNUAL WORTH COMPARISON Calculator***, an Excel spreadsheet created by the author and located in the *Book Templates* file. We use it to perform an annual worth comparison of two or more cost alternatives with lives that equal or *exceed* the service time. The cash flows of each alternative consist of a purchase price, a salvage or **market value**,[3] and *any* combination of the three basic series. *The spreadsheet automatically calculates the values of the annual worths and their sums, for arbitrary values of the INPUT variables when g% and i% are different.*

To use the ***ANNUAL WORTH COMPARISON Calculator*** in a least-cost study, make a copy of it and enter the values of the variables in the *INPUT* columns. NOTE: As soon as you change the value of an alternative's input variable, the spreadsheet automatically changes the values of the corresponding worths.

[2] The worths of the cash flow elements in a worth table are always rounded to the nearest dollar and *then* added. Occasionally, this will result in *negligible* round-off errors in the annual worths.
[3] *When the service time is less than the life of an alternative, we use its market value in a least-cost study.* The **market value** *MV* of a piece of equipment, such as a compressor or a heat exchanger, is its value in the used equipment market. It is generally greater than the equipment's salvage value.

	A	B	C	D	E	F	G	H
1								
2		***ANNUAL WORTH COMPARISON Calculator***						
3			***Input:***			***Solution: Annual Worths and Sums***		
4			***W***	***X***	***Y***	***W***	***X***	***Y***
5		***P***	\$75,000	\$80,000	\$85,000	\$17,889	\$19,082	\$20,274
6		***S ⊕ MV***	(\$3,500)	(\$4,000)	(\$4,500)	(\$135)	(\$154)	(\$173)
7		***A***	\$3,350	\$3,000	\$2,200	\$3,350	\$3,000	\$2,200
8		***G***	\$950	\$0	\$0	\$2,920	\$0	\$0
9		***A_1***	\$0	\$750	\$460	\$0	\$730	\$448
10		***g%***	0%	5%	5%			
11		***$A_{1,net}$***	0	1	1			
12		***r*%***	20%		***ΣAW***	**\$24,024**	**\$22,658**	**\$22,749**
13		***ST***	10					

Spreadsheet 4.1 ***ANNUAL WORTH COMPARISON Calculator*** NOTES: (1) Alternatives *X* and *Y* have the *same* economic data and annual worths as compressors *X* and *Y* in Example 4.1. (2) The symbol ***S ⊕ MV*** in cell B6 means ***salvage value*** *S or* ***market value*** *MV*. (3) Placing a thick black border on the uniform worth cell of the least-cost alternative quickly identifies it.

4.2.3 Using the *ANNUAL WORTH COMPARISON* Calculator

Specifying the Annual Worth or Net Annual Worth of a Geometric Gradient Series

When a value of $A_{1,\text{net}}$ in array C11:F11 is 0, the annual worth of the geometric gradient series is calculated. When a value of $A_{1,\text{net}}$ in array C11:F11 is 1, the net annual worth of the geometric gradient series is calculated. For example, in Spreadsheet 4.1, the net annual worths of the geometric gradient series of alternatives *X* and *Y* are calculated.

Adding More Cost Alternatives

If you want to *add* more cost alternatives to the ***ANNUAL WORTH COMPARISON Calculator***, make a copy of it and do the following:

1. *Select* one or more *adjacent* columns to the *right* of the E column > *Right Click* the selected column(s) > *Click* **Insert**.
2. *Enter* the name(s) of the cost alternative(s) in Row 4 of the inserted *INPUT* column(s).
3. *Enter* the values of *P*, *S ⊕ MV*, *A*, *G*, A_1, *g%*, and $A_{1,\text{net}}$ in Rows 5 through 11 of the inserted *INPUT* column(s).
4. *Paste* the *original* array H5:H13 into the *empty SOLUTION* column(s) to the *right* of the *original* H column to evaluate the solution cells of the added cost alternative(s) > *Enter* the name(s) in Row 4 of the *SOLUTION* column(s).
5. *Merge* the cells in each title row.

4.2.4 Component Replacement Costs

Sometimes an alternative has a *costly* component that must be replaced during its life. For example, the lining of a furnace, the catalyst of a chemical reactor, and the surface of a road are often replaced one or more times during their alternative's lives. A **component replacement cost** at the end of the kth year C_k is defined as

$$C_k \equiv C_0(1+f)^k, \tag{4-3}$$

where C_0 is the cost at the *beginning* of the service time, and f is the annual rate at which it increases. It depends on the prevailing economic climate, and a range of $0.02 < f < 0.05$ is typical. Example 4.2 illustrates the use of Equation (4-3).

EXAMPLE 4.2 Finding the Least-Cost Reactor: An Annual Worth Comparison with Component Replacement Costs When $L > ST = 10$ a (*AWC*-Class)

Problem: A chemical company is considering the construction of a plant to produce cumene, which is a paint thinner and a component of high octane fuels. Two American process engineering companies have submitted bids to supply their proprietary catalytic reactors to carry out the reaction. As with the compressors for a chemical plant, the MARR% is 20% per year and the service time is 10 years.

The cost of reactor X *and* its catalyst is \$165,000, and the *estimated* market value of the reactor at the end of the service time is \$11,800. The initial cost of the catalyst is \$34,000. It deteriorates with time and must be replaced every four years. The energy cost is \$4,000 in the first year, and it is *expected* to increase by \$250 in each subsequent year of the reactor's life (12 a).

The cost of Reactor Z *and* its catalyst is \$200,000, and the *estimated* market value of the reactor at the end of the service time is \$25,400. The initial cost of the catalyst is \$22,000. It too deteriorates with time and must be replaced every five years. The energy cost is \$1,200 in the first year, and it is *expected* to increase by \$150 in each subsequent year in the reactor's life (15 a). *Use an annual worth comparison to find the least-cost reactor.* NOTE: *All replacement costs increase at a rate of 3% per year.*

Solution: Table 4.3 lists the economic data for reactors X and Z.

TABLE 4.3 Economic Data for the Reactors

DATA TYPE	X	Z
Reactor Purchase Price: P	\$165,000	\$200,000
Reactor Life: L	12 a	15 a
Market Value: MV	\$11,800	\$25,400
Initial Catalyst Cost: C_0	\$34,000	\$22,000
Catalyst Life: l	4 a	5 a
Annual Percentage Increase of Catalyst Cost: f%	3%	3%
First Year Energy Cost: A	\$4,000	\$1,200
Annual Energy Cost Increment: G	\$250	\$150
Service Time: ST	10 a	

Book Convention: *Tabulating economic data that is presented in a narrative form clearly reveals the differences between the alternatives and minimizes the chance that one or more incorrect data values will be used in a least-cost study. This convention is followed throughout the book.*

Solution: Table 4.4 is an annual worth table that displays the annual worths of the cash flow elements of reactors X and Z when the MARR% is 20% per year.

TABLE 4.4 Comparing the Costs of Reactors X and Z: $L > ST = 10$ a and $r^*\% = 20\%$ per Year

WORTH TYPE	REACTOR ANNUAL WORTH TABLE	
	X	Z
AW_P	$\$165{,}000\underbrace{(A/P, 20\%, 10)}_{0.23852} = \$39{,}356$	$\$200{,}000\underbrace{(A/P, 20\%, 10)}_{0.23852} = \$47{,}704$
AW_{MV}	$\$11{,}800\underbrace{(A/F, 20\%, 10)}_{0.03852} = (\$455)$	$\$25{,}400\underbrace{(A/F, 20\%, 10)}_{0.03852} = (\$978)$
AW_{C_k}	$\$34{,}000\left[\begin{array}{l}(1+0.03)^4\underbrace{(P/F, 20\%, 4)}_{0.48225} + \\ +(1+0.03)^8\underbrace{(P/F, 20\%, 8)}_{0.23257}\end{array}\right] \times \underbrace{(A/P, 20\%, 10)}_{0.23852} = \$6{,}791$	$\$22{,}000(1+0.03)^5\underbrace{(P/F, 20\%, 5)}_{0.40188} \times \underbrace{(A/P, 20\%, 10)}_{0.23852} = \$2{,}445$
AW_A	\$4,000	\$1,200
AW_G	$\$250\underbrace{(A/G, 20\%, 10)}_{3.0739} = \$768$	$\$150\underbrace{(A/G, 20\%, 10)}_{3.0739} = \$461$
ΣAW	**\$50,460**	**\$50,832**

Reactor X is the least-cost reactor.

4.2.5 Adding Component Replacement Costs to the *ANNUAL WORTH COMPARISON* Calculator

Spreadsheet 4.2 shows the ***ANNUAL WORTH COMPARISON Calculator with COMPONENT REPLACEMENT*** spreadsheet in the *Book Templates* file. It is the ***ANNUAL WORTH COMPARISON Calculator*** with the addition of *new* rows 7, 8, 9, and 10 for the initial cost of a component C_0, its life l (measured in annual interest periods), and the annual *percentage* increase of its cost $f\%$. The symbol $\#_{max} = 2$ specifies the maximum number of component replacements in this spreadsheet: *one* replacement (value 0) or *two* replacements (value 1) in the ST interest periods. *If an alternative has no replacement costs, enter zero for C_0, l, $f\%$, and $\#_{max}$.*

Computation Notes: *(1) Element annual worths are calculated with annual worth factor* ***equations*** *and rounded to the dollar.*

	A	B	C	D	E	F	G	H
1								
2		***ANNUAL WORTH COMPARISON Calculator with COMPONENT REPLACEMENT***						
3		***Input:***				***Solution: Annual Worths and Sums***		
4			***X***	***Y***	***Z***	***X***	***Y***	***Z***
5		***P***	\$165,000	\$185,000	\$200,000	\$39,356	\$44,127	\$47,705
6		***S ⊕ MV***	(\$11,800)	(\$23,000)	(\$25,400)	(\$455)	(\$886)	(\$978)
7		C_0	\$34,000	\$28,000	\$22,000	\$6,791	\$3,265	\$2,445
8		***l***	4	5	5			
9		***f%***	3%	4%	3%			
10		$\#_{max} = 2$	1	0	0			
11		***A***	\$4,000	\$1,800	\$1,200	\$4,000	\$1,800	\$1,200
12		***G***	\$250	\$175	\$150	\$768	\$538	\$461
13		A_1	\$0	\$300	\$0	\$0	\$281	\$0
14		***g%***	0%	4%	0%			
15		$A_{1,net}$	0	1	0			
16		***r*%***	20.00%		***ΣAW***	**\$50,460**	**\$49,125**	**\$50,833**
17		***ST***	10					

Spreadsheet 4.2 ***ANNUAL WORTH COMPARISON Calculator with COMPONENT REPLACEMENT*** NOTE: This calculates the annual worths of reactors *X* and *Z* in Example 4.2 and a reactor *Y* that is the least-cost reactor.

*(2) A **manual** calculation of an element annual worth uses a factor **value** (with five decimal places) from the **12-FACTOR Calculator** and rounds it to the dollar. Occasionally this value may be one dollar smaller than the factor equation value.*

4.2.6 Effect of the MARR on a Least-Cost Study

Spreadsheet 4.2 shows that reactor *Y* is the least-cost reactor. Let's say that you present this result to your manager in industry. He/she may ask you the following reasonable question: *What happens if we change the MARR?* The following two small paragraphs provide the answer.

Effect of the MARR on the Annualizing Factors

The factors that annualize *S* ⊕ *MV*, *G*, and A_1 convert them to equivalent annual series. As r^* *increases*, the metrics of the series must *decrease* to maintain the equivalence (you can confirm this with the ***12-FACTOR Calculator***). Eventually these metrics become negligible and have no effect on the conclusion of the least-cost study.

Effect of the MARR on the Capital Recovery Factors

As the value of r^* increases, the capital recovery factor increases. Hence, ***the alternative with the smallest purchase price P always becomes or remains the least-cost alternative.*** Spreadsheet 4.3 illustrates this important conclusion.

	A	B	C	D	E	F	G	H
1								
2		***ANNUAL WORTH COMPARISON Calculator with COMPONENT REPLACEMENT***						
3		***Input***				***Solution: Annual Worths and Sums***		
4			***X***	***Y***	***Z***	***X***	***Y***	***Z***
5		***P***	$165,000	$185,000	$200,000	$47,212	$52,934	$57,226
6		***S ⊕ MV***	($11,800)	($23,000)	($25,400)	($343)	($668)	($737)
7		***C_0***	$34,000	$28,000	$22,000	$6,360	$3,105	$2,324
8		***l***	4	5	5			
9		***f%***	3%	4%	3%			
10		**$\#_{max}$ = 2**	1	0	0			
11		***A***	$4,000	$1,800	$1,200	$4,000	$1,800	$1,200
12		***G***	$250	$175	$150	$690	$483	$414
13		***A_1***	$0	$300	$0	$0	$268	$0
14		***g%***	0%	4%	0%			
15		***$A_{1,net}$***	0	1	0			
16		***r*%***	25.71%		**ΣAW**	**$57,919**	**$57,922**	**$60,427**
17		**ST**	10					

Spreadsheet 4.3 ***ANNUAL WORTH COMPARISON Calculator with COMPONENT REPLACEMENT*** NOTE: When $r^*\%$ is 25.71% or greater, the *least expensive* reactor *X* is the *least-cost* alternative.

4.3 INDEFINITE SERVICE TIME

4.3.1 Least Common Multiple

Table 4.5 lists the economic data for two 3,200 ft^2 shell and tube heat exchangers that are being compared for an indefinite service time in a power plant. The more expensive stainless steel exchanger lasts longer because of its excellent corrosion resistance. This also accounts for the lower annual maintenance cost for removing

TABLE 4.5 Economic Data for Heat Exchangers *X* and *Y*

DATA TYPE	*X*	Y
Material of Construction	Carbon Steel	316 Stainless Steel
Purchase Price: *P*	$26,000	$45,000
Salvage Value: *S*	$1,000	$4,500
Annual Maintenance Cost: *A*	$2,500	$300
First Year Energy Cost: A_1	$700	$500
Annual Rate of Change of Energy Costs: *g%*	3%	3%
Life: *L*	5 a	10 a

corrosion scale from the interior and exterior surfaces of the tubes. Since exchanger Y has cleaner shell and tube surfaces, its annual energy cost for pumping the shell and tube fluids through it is also lower than that of exchanger X.

Since the service time is indefinite, the simplest way to compare these exchangers is to *assume* that it equals the **least common multiple** *LCM* of the lives of the exchangers. That is, for *any two* alternatives X and Y,

$$\boxed{ST = LCM \equiv \begin{cases} m_X L_X \\ m_Y L_Y, \end{cases}} \tag{4-4}$$

where m_X and m_Y are the *smallest* integers that satisfy Equation (4-4). Thus, for exchangers X and Y the service time is given by

$$ST = LCM = \begin{cases} m_X L_X = 2 \times 5 \text{ a} = 10 \text{ a} \\ m_Y L_Y = 1 \times 10 \text{ a} = 10 \text{ a}. \end{cases}$$

The 10-year service time contains *two* lives of exchanger X and *one* life of exchanger Y. This means that exchanger X is replaced *one* time.

4.3.2 Repeatability Assumption

Definition

The practical application of an *LCM* service time is based on the **Repeatability Assumption** (RA):

Repeatability Assumption
The cash flows on each life of an alternative are *identical*.

Figure 4.1 shows the cash flows of exchanger X on its two lives when the RA is valid. Since the cash flows on each life are identical, the annual worth $AW_{X,1}$ of the

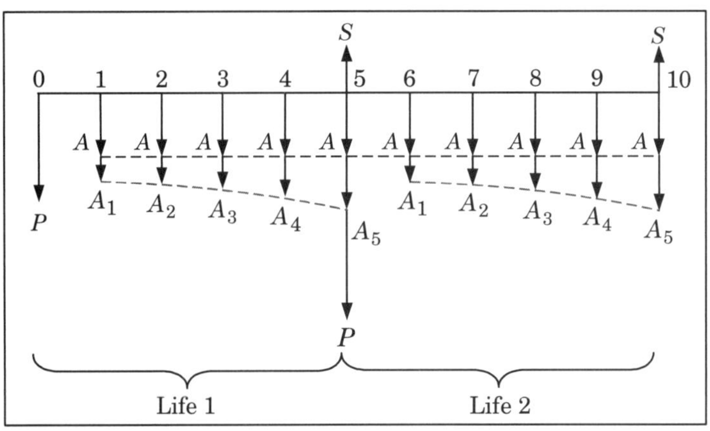

Figure 4.1 **Cash Flows of Exchanger X on the 10-Year Service Time**
NOTES: (1) *The cash flows on each life are identical because of the RA.* (2) The *first* salvage value and the replacement cost occur at the *end* of the first life.

cash flows on the first life equals that on the second life $AW_{X,2}$. Hence, ***the annual worth of the cash flows of exchanger X on the 10-year service time AW_X equals $AW_{X,1}$***. We can extend this result to any alternative Z with m_z lives:

$$\boxed{AW_Z = AW_{Z,1} = AW_{Z,2} = \cdots = AW_{Z,m_Z}.} \tag{4-5}$$

Table 4.6 is an LCM annual worth table based on the RA. It compares the annual worths of exchangers X and Y when the MARR% is 11% per year. Even though it has the larger purchase price, *Exchanger Y is the least-cost alternative.*

TABLE 4.6 Annual Worth Comparison of Exchangers X and Y Based on the RA: $ST = LCM = 10$ a and $r^*\% = 11\%$ per Year

WORTH TYPE	EXCHANGER ANNUAL WORTH TABLE	
	X	Y
AW_P	$\$26{,}000\underbrace{(A/P,11\%,5)}_{0.27057} = \$7{,}035$	$\$45{,}000\underbrace{(A/P,11\%,10)}_{0.16980} = \$7{,}641$
AW_S	$\$1{,}000\underbrace{(A/F,11\%,5)}_{0.16057} = (\$161)$	$(\$4{,}500)\underbrace{(A/F,11\%,10)}_{0.05980} = (\$269)$
AW_A	\$2,500	\$300
AW_{A_1}	$\$700\underbrace{(A/A_1,3\%,11\%,5)}_{1.05533} = \$739$	$\$500\underbrace{(A/A_1,3\%,11\%,10)}_{1.11792} = \$559$
ΣAW	**\$10,113**	**\$8,231**

4.3.3 Comparison of Specified & *LCM* Service Times

The RA states that the cash flows of an alternative are identical on each of its lives in an LCM service time. Hence, it neglects changes in the cash flows of an alternative's replacement costs and its component replacement costs. Example 4.3 illustrates this and compares the result with that of Example 4.2, which compares the same alternatives on a specified service time.

EXAMPLE 4.3 Finding the Least-Cost Reactor: An Annual Worth Comparison with Replacement Costs and an *LCM* Service Time (*AWC*-Class)

Problem: A project engineer for the company that is considering the cumene plant investment knows that the company has occasionally used an LCM service time to compare cost alternatives. Table 4.7 is the economic data for a least-cost comparison of the cumene reactors on their LCM service time. *If the MARR% is still 20% per year and the service time is indefinite, which reactor is the least-cost alternative?* NOTE: All replacement costs increase at a rate of 3% per year.

Solution: Table 4.7 shows that the economic data for reactors X and Y is almost identical to that in Table 4.3 except that salvage values replace market values and $ST = LCM = 60$ a.

TABLE 4.7 Economic Data for the Reactors

DATA TYPE	*X*	Y
Reactor Purchase Price: *P*	$165,000	$200,000
Reactor Life: *L*	12 a	15 a
Salvage Value: *S*	$7,000	$15,000
Initial Catalyst Cost: C_0	$34,000	$22,000
Catalyst Life: *l*	4 a	5 a
Annual Percentage Increase of Catalyst Cost: *f*%	3%	3%
First Year Energy Cost: *A*	$4,000	$1,200
Annual Energy Cost Increment: *G*	$250	$150
Service Time: *LCM*	60 a	

Table 4.8 compares the annual worths of the cash flow elements of reactors *X* and *Y* when the MARR% is 20% per year. *NOTE: The present worth of a catalyst replacement cost is annualized on its reactor life to obtain its annual worth.*

TABLE 4.8 Annual Worth Comparison of Reactors *X* and *Y*: *ST* = *LCM* = 60 a and *r% = 20% per Year**

WORTH TYPE	REACTOR ANNUAL WORTH TABLE	
	X	*Y*
AW_P	$\$165{,}000\underbrace{(A/P,20\%,12)}_{0.22526} = \$37{,}168$	$\$200{,}000\underbrace{(A/P,20\%,15)}_{0.21388} = \$42{,}776$
AW_S	$\$7{,}000\underbrace{(A/F,20\%,12)}_{0.02526} = (\$177)$	$(\$15{,}000)\underbrace{(A/F,20\%,15)}_{0.01388} = (\$208)$
AW_{C_k}	$\$34{,}000\left[\begin{matrix}(1+0.03)^4\underbrace{(P/F,20\%,4)}_{0.48225}+\\+(1+0.03)^8\underbrace{(P/F,20\%,8)}_{0.23257}\end{matrix}\right]\times$ $\times\underbrace{(A/P,20\%,12)}_{0.22526} = \$6{,}413$	$\$22{,}000\left[\begin{matrix}(1+0.03)^5\underbrace{(P/F,20\%,5)}_{0.40188}+\\+(1+0.03)^{10}\underbrace{(P/F,20\%,10)}_{0.16151}\end{matrix}\right]\times$ $\times\underbrace{(A/P,20\%,15)}_{0.21388} = \$3{,}214$
AW_A	$4,000	$1,200
AW_G	$\$250\underbrace{(A/G,20\%,12)}_{3.48410} = \$871$	$\$150\underbrace{(A/G,20\%,15)}_{3.95884} = \$594$
ΣAW	**$48,275**	**$47,576**

Even though it costs more, ***reactor Y is now the least-cost alternative!***

Example 4.3 shows how to calculate the annual worths of the two reactors with an *LCM* service time. ***But the conclusion of this study is the opposite of that in the specified service time study in Example 4.2.*** The reason for this is the annualizing of most of the costs over 12 and 15years versus the specified 10 years used in Example 4.2.

4.3.4 Disadvantages of the *LCM* Service Time

Let's summarize the disadvantages of using an *LCM* service time.

1. The calculated annual worths are less accurate than those calculated with a specified service time because of the Repeatability Assumption.
2. *The least-cost alternative calculated with an LCM service time may not be the same as the one calculated with a specified service time!*
3. We can't conveniently compare three or more alternatives because an *LCM* service time rarely exists.
4. An *LCM* service time normally doesn't equal the life of a *prospective* investment that incorporates the least-cost alternative. For example, it would be *unreasonable* for a project manager to incorporate the 60-year *LCM* annual cost of the reactors into a *prediction* of the IRR% of a proposed cumene plant with an *assumed* 10-year operating life.
5. When an *LCM* service time contains many lives of the two alternatives, their incremental investment is usually a *non-simple* investment.

Computation Note: *To eliminate these disadvantages, a specified service time is used in all subsequent least-cost studies in the book.*

4.4 OTHER COST ALTERNATIVES

4.4.1 Extra Capacity Versus Staged Construction

Capacity and Demand

Capacity is the number and unit of a **physical quantity**, such as area and volume, provided by an alternative, and **need** is the task that it performs. **Demand** is synonymous with need. Table 4.9 lists examples of the capacities and demands of compressors, heat exchangers, and single-family houses.

TABLE 4.9 Capacities and Demands of Three of Alternatives

COST ALTERNATIVE	CAPACITY[1]	DEMAND
Compressor	Power[2]: 100 hp	Increase the pressure of 50,000 $lb{\cdot}h^{-1}$ of ambient air to 75 psi.
Heat Exchanger	Area: 3,200 ft^2	Transfer 2×10^6 $BTU{\cdot}h^{-1}$ to 35,000 $lb{\cdot}h^{-1}$ of boiler feedwater.
Single-Family House	Volume: 20,000 ft^3	Provide a place for a family to sleep, bathe, eat, and entertain.

[1] Power, area, and volume are the **physical quantities** provided by a compressor, heat exchanger, and a single-family house, respectively.
[2] The symbol for power measured in horsepower (550 $ft{\cdot}lbf{\cdot}s^{-1}$) is "hp."

Definitions of Extra Capacity and Staged Construction

In the previous present worth comparisons of two compressors, reactors, and heat exchangers, it is tacitly *assumed* that *the capacity of each alternative can only satisfy a present demand.* But a demand may increase with time and a future purchase of an alternative will be required. When an *alternative* can satisfy a present *and* future demand, we call it an **extra capacity (EC) alternative**. A common economic comparison is between the construction of an extra capacity object now and the **staged construction** of a smaller object now and in the future.

EXAMPLE 4.4 Finding the Least-Cost Alternative When One Is an Extra Capacity Alternative: An Annual Worth Comparison with a Specified Service Time and Staged Construction (*AWC*-Class)

Problem: A town wants to build a new elementary school because of the increasing population of its young marrieds. A consultant ***estimates*** that an elementary school for 100 children will be adequate for the next 6 years. However, at the *end* of the 6th year *another* one will be required for an *additional* 100 children. An EC alternative *Y* to the two schools (alternative *X*) is *one* larger elementary school that can serve 100 children *now* and 200 in the *future*. Table 4.10 lists the economic data for the alternatives. *Find the least-cost alternative when (1) the MARR% is 8% per year,*[4] *(2) ST = 15 a, (3) construction costs increase by 3% per year, and (4) the market values of the buildings can be neglected.*

Solution: Table 4.11 compares the annual worths of the cash flow elements of the *X* and *Y alternatives*. As you can see, *the extra capacity school Y is significantly less expensive to build and operate than the two smaller X schools.*

TABLE 4.10 Economic Data for Elementary Schools *X* (Two Schools) and *Y* (One School)

DATA TYPE	X	*Y*
School Price for X_{1st} and *Y*: *P*	$\$1.2\times10^6$	$\$2\times10^6$
School Price for X_{2nd} at TP_6: $P_{2nd} = \$1.2\times10^6\times(1.03)^6$	$\$1.433\times10^6$	—
First Year Operating Cost for X_{1st} and *Y*: *A*	\$130,000	\$180,000
Annual Operating Cost Increment for X_{1st} and *Y*: *G*	\$3,750	\$6,000
First Year Operating Cost for X_{2nd} at TP_7: A_{2nd}	\$152,500	—
Annual Operating Cost Increment for X_{2nd} at TP_8: G_{2nd}	\$3,750	—

TABLE 4.11 Annual Worth Comparison of Elementary Schools *X* (Two Schools) and *Y* (One School): *ST* = 15 a & $r^*\%$ = 8% per Year[1]

WORTH TYPE	SCHOOL ANNUAL WORTH TABLE	
	X	*Y*
AW_P	$\$1.2\times10^6\underbrace{(A/P,8\%,15)}_{0.11683} = \$140{,}196$	$\$2\times10^6\underbrace{(A/P,8\%,15)}_{0.11683} = \$233{,}660$
AW_{P_2}	$\$1.433\times10^6\underbrace{(P/F,8\%,6)}_{0.63017}\underbrace{(A/P,8\%,15)}_{0.11683} = \$105{,}501$	—
AW_A	\$130,000	\$180,000

[4]The small MARR% represents the interest rate of the bonds that will be issued to finance the least-cost alternative.

TABLE 4.11 Annual Worth Comparison of Elementary Schools *X* (Two Schools) and *Y* (One School): *ST* = 15 a & *r% = 8% per Year[1] *(continued)***

Worth Type	School Annual Worth Table	
	X	*Y*
AW_G	$\$3{,}750\underbrace{(A/G,8\%,15)}_{5.5945} = \$20{,}979$	$\$6{,}000\underbrace{(A/G,8\%,15)}_{5.5945} = \$33{,}567$
AW_{A_2}	$\$152{,}500\underbrace{(F/A,8\%,9)}_{12.488}\underbrace{(A/F,8\%,15)}_{0.036830} = \$70{,}140$	—
AW_{G_2}	$\$3{,}750\underbrace{(F/G,8\%,9)}_{43.594}\underbrace{(A/F,8\%,15)}_{0.036830} = \$6{,}021$	—
ΣAW	**$472,837**	**$447,227**

[1] ***The present time for the second uniform and gradient series is*** TP_6.

The least-cost alternative is the extra capacity elementary school Y.

Power Law Cost Scale-up Rule

The capacity, measured in ft^2, of the *Y* school is twice that of one *X* school. However, Table 4.10 shows that the cost of the *Y* school ($\$2 \times 10^6$) is significantly *less* than twice the cost of one *X* school ($2 \times \$1.2 \times 10^6 = \$2.4 \times 10^6$). This illustrates the **Power Law Cost Scale-up Rule**[5]:

TABLE 4.12

Power Law Cost Scale-up Rule
The ratio of the costs of a larger object and a smaller object of the same type varies as the ratio of their capacities raised to a positive power that is less than 1. That is, $$C_2 = C_1\left(\frac{K_2}{K_1}\right)^x. \quad (4\text{-}6)$$ where C_1 and C_2 are the costs of the smaller and larger objects, respectively, K_1 and K_2 are their capacities, and $0 < x < 1$.

Usage Notes: *(1) The value of x depends on the class of objects, such as heat exchangers or compressors.*
(2) It also depends on the type of object within a class; for example, some types of heat exchangers are double pipe, shell & tube, and plate.

[5] Garnett, D.I. and Patience, G.S, *Why Do Scale-up Power Laws Work?*, Chemical Engineering Progress, August 1993, pp. 76-78.

4.4.2 Dissimilar Alternatives

We have previously compared the annual worths of the cash flows of *similar* alternatives, such as three reactors or two small elementary schools and a larger one. However, we must often compare the costs of **dissimilar alternatives**. Here are two examples of common dissimilar alternatives and their cash flow elements:

- *Make Objects* VERSUS *Buy Objects*: equipment cost & annual operating costs VERSUS annual purchasing costs
- *Labor* VERSUS *Equipment:* annual labor cost VERSUS equipment cost & annual operating costs

EXAMPLE 4.5 Finding the Least-Cost of Engraving iPods: An Annual Worth *Make* or *Buy* Comparison with a Specified Service Time (*AWC*-Class)

Problem: A major retailer of iPods currently sends them out for custom engraving at a cost \$17.50 per iPod for two lines of text with one font. This is the *Buy* alternative. A 20 watt CO_2 laser engraving machine that can engrave text and graphics costs \$27,000, and its operating cost is \$9.50 per iPod. This is the *Make* alternative. Table 4.13 lists the retailer's *expected* engraved iPod sales *ES* for the next five years and the corresponding custom engraving costs and the operating costs of the engraving machine, which has a *negligible* salvage value:

TABLE 4.13 Effect of Expected Engraved iPod Sales on *Make* & *Buy* Costs

Expected Engraved iPod Sales *ES*	1,200	1,350	1,320	1,270	1,500
Year k	1	2	3	4	5
Buy Costs: \$17.50 × *ES*	\$21,000	\$23,625	\$23,100	\$22,225	\$26,250
Make Operating Costs: \$9.50 × *ES*	\$11,400	\$12,825	\$12,540	\$12,065	\$14,250

If the MARR% is 10% per year, should the retailer continue sending out the iPods for engraving?

Solution: Table 4.14 compares the annual worths of the cash flow elements of the *Make* and *Buy* alternatives.

TABLE 4.14 Annual Worth Comparison of the *Make* and *Buy* iPod Engraving Alternatives: $ST = 5$ a & $r^{*}\% = 10\%$ per Year

WORTH TYPE	MAKE OR BUY ANNUAL WORTH TABLE	
	MAKE	BUY
AW_P	$\$27{,}000\underbrace{(A/P,10\%,5)}_{0.26380} = \$7{,}123$	—
AW_{C_1}	$\$11{,}400\underbrace{(P/F,10\%,1)}_{0.90909}\underbrace{(A/P,10\%,5)}_{0.26380} = \$2{,}734$	$\$21{,}000\underbrace{(P/F,10\%,1)}_{0.90909}\underbrace{(A/P,10\%,5)}_{0.26380} = \$5{,}036$

TABLE 4.14 Annual Worth Comparison of the *Make* and *Buy* iPod Engraving Alternatives: $ST = 5$ a & $r^*\% = 10\%$ per Year *(continued)*

WORTH TYPE	MAKE OR BUY ANNUAL WORTH TABLE	
	MAKE	*BUY*
AW_{C_2}	$\$12{,}825\underbrace{(P/F,10\%,2)}_{0.82645}\underbrace{(A/P,10\%,5)}_{0.26380} = \$2{,}796$	$\$23{,}625\underbrace{(P/F,10\%,2)}_{0.82645}\underbrace{(A/P,10\%,5)}_{0.26380} = \$5{,}151$
AW_{C_3}	$\$12{,}540\underbrace{(P/F,10\%,3)}_{0.75131}\underbrace{(A/P,10\%,5)}_{0.26380} = \$2{,}485$	$\$23{,}100\underbrace{(P/F,10\%,3)}_{0.75131}\underbrace{(A/P,10\%,5)}_{0.26380} = \$4{,}578$
AW_{C_4}	$\$12{,}065\underbrace{(P/F,10\%,4)}_{0.68301}\underbrace{(A/P,10\%,5)}_{0.26380} = \$2{,}174$	$\$22{,}225\underbrace{(P/F,10\%,4)}_{0.68301}\underbrace{(A/P,10\%,5)}_{0.26380} = \$4{,}004$
AW_{C_5}	$\$14{,}250\underbrace{(P/F,10\%,5)}_{0.62092}\underbrace{(A/P,10\%,5)}_{0.26380} = \$2{,}334$	$\$26{,}250\underbrace{(P/F,10\%,5)}_{0.62092}\underbrace{(A/P,10\%,5)}_{0.26380} = \$4{,}300$
ΣAW	$19,646	$23,069

The *Make* alternative saves the retailer $3,423 per year. Hence, *the retailer should buy the engraving machine and* **stop** *sending the iPods out for engraving.*

4.4.3 Irregular Alternatives

Irregular alternatives have numerous *unrelated* cash flows on the service time. For example, the *Make* and *Buy* alternatives of Example 4.5 are also irregular alternatives. One way to calculate the *annual* worth of an irregular alternative is to

1. *Annualize* the present worth of *each* of its cash flows.
2. Add the annual worths calculated in Step 1 to obtain the annual worth of the alternative.

Example 4.5 illustrates this cumbersome approach.

A simpler way to calculate the *annual* worths of irregular alternatives is the **Consistency Equation Method** defined by **Step-by-Step 4.1**:

Calculating the Annual Worth of an Irregular Alternative Step-by-Step 4.1
(1) Calculate the present worth of an irregular alternative X.
(2) Calculate its annual worth with the Consistency Equation 4-1: $AW_X = PW_X(A/P, r^*, ST)$.

EXAMPLE 4.6 Using the Consistency Equation Method to Find the Least-Cost of Engraving iPods (*AWC*-Class)

Problem: The following table lists the annual engraving costs (the Buy alternative) and the annual operating costs of the $27,000 laser engraving machine (the Make alternative) in Example 4.5. Use Step-by-Step 4.1 to compare the annual worths of the Make and Buy alternatives when the MARR% is 10% per year, the value used in Example 4.5.

FUTURE YEAR k	1	2	3	4	5
ENGRAVING COSTS C_k	$21,000	$23,625	$23,100	$22,225	$26,250
OPERATING COSTS C_k	$11,400	$12,825	$12,540	$12,065	$14,250

Solution: ***Step1: Calculate the Present Worths of the Alternatives***

Table 4.15 compares the present worths of the *Make* and *Buy* alternatives:

TABLE 4.15 Present Worth Comparison of the *Make (M)* and *Buy (B)* iPod Engraving Alternatives: $ST = 5$ a, and $r^{*}\% = 10\%$ per Year

WORTH TYPE	MAKE OR BUY ANNUAL WORTH TABLE	
	MAKE	*BUY*
PW_P	$27,000	
PW_{C_1}	$\$11{,}400\underbrace{(P/F,10\%,1)}_{0.90909}== \$10{,}364$	$\$21{,}000\underbrace{(P/F,10\%,1)}_{0.90909} = \$19{,}091$
PW_{C_2}	$\$12{,}825\underbrace{(P/F,10\%,2)}_{0.82645} = \$10{,}599$	$\$23{,}625\underbrace{(P/F,10\%,2)}_{0.82645} = \$19{,}525$
PW_{C_3}	$\$12{,}540\underbrace{(P/F,10\%,3)}_{0.75131} = \$9{,}421$	$\$23{,}100\underbrace{(P/F,10\%,3)}_{0.75131} = \$17{,}355$
PW_{C_4}	$\$12{,}065\underbrace{(P/F,10\%,4)}_{0.68301} = \$8{,}241$	$\$22,225\underbrace{(P/F,10\%,4)}_{0.68301} = \$15{,}180$
PW_{C_5}	$\$14{,}250\underbrace{(P/F,10\%,5)}_{0.62092} = \$8{,}848$	$\$26{,}250\underbrace{(P/F,10\%,5)}_{0.62092} = \$16{,}299$
ΣPW	**$74,473**	**$87,450**

Step 2: Calculate the Annual Worths of the Alternatives

It follows from the results in Table 4.15 and Step-by-Step 4.1 that the annual worths of the alternatives are

$$AW_M = \underbrace{PW_M}_{\$74{,}473}\underbrace{(A/P,10\%,5)}_{0.26380} = \underline{\underline{\$19{,}646}}$$

$$AW_B = \underbrace{PW_B}_{\$87{,}450}\underbrace{(A/P,10\%,5)}_{0.26380} = \underline{\underline{\$23{,}069}}.$$

As expected, *these annual worths are identical to those obtained in Example 4.5.* However, calculating the present worths of the alternatives and multiplying them by the *common* capital recovery factor is much simpler than calculating the annual worths of the cash flow elements.

4.5 PERPETUAL SERVICE TIME

4.5.1 Definition

Our national infrastructure contains structures, such as dams, aqueducts, bridges, tunnels, and highways, that will essentially last "forever" if they are properly maintained. That is, they are designed for a **perpetual service time** ST_{∞}. To compare alternatives designed for a perpetual service time, we will need some or all of the annual worth limits of a present cash flow, a gradient series, and a geometric gradient series as the number of interest periods approaches infinity.

4.5.2 Three Annual Worth Limits

Present Cash Flow

The definition of the **limit** of the **annual worth of a present cash flow** $AW_{P,\infty}$ is obtained from Equation (2-52) in Subsection 2.6.3:

$$\boxed{AW_{P,\infty} \equiv P \lim_{N\to\infty} \frac{i}{1-(1+i)^{-N}} = Pi.} \tag{4-7}$$

Gradient Series

The definition of the **limit** of the **annual worth of a gradient series** $AW_{G,\infty}$ is obtained from Equation (2-87) in Subsection 2.7.3. However, this limit yields an indeterminate result:

$$AW_{G,\infty} \equiv G \lim_{N\to\infty}\left[\frac{1}{i} - \frac{N}{(1+i)^N - 1}\right] = G\left[\frac{1}{i} - \frac{\infty}{(1+i)^{\infty}}\right].$$

Hence, we apply L'Hôpital's Rule to calculate the limit:

$$AW_{G,\infty} = \frac{G}{i} - G \lim_{N\to\infty}\left[\frac{\frac{d}{dN}N}{\frac{d}{dN}\left\{(1+i)^N - 1\right\}}\right]$$

$$AW_{G,\infty} = \frac{G}{i} - \underbrace{G \lim_{N\to\infty}\left[\frac{1}{N(1+i)^{N-1}}\right]}_{0}.$$

Thus,

$$AW_{G,\infty} = \frac{G}{i}. \tag{4-8}$$

Geometric Gradient Series

The definition of the **limit** of the **annual worth of a geometric gradient series** $AW_{A_1,\infty}$ is obtained from Equation (2-103) in Subsection 2.8.4:

$$AW_{A_1,\infty} = A_1 \lim_{N\to\infty}\left[\frac{i}{i-g}\,\frac{1-\left(\frac{1+g}{1+i}\right)^N}{1-(1+i)^{-N}}\right] \qquad g \neq i.$$

When $i > g$, the numerator of the second quotient converges to 1, and its denominator always converges to 1. Hence,

$$AW_{A_1,\infty} = A_1\frac{i}{i-g} \qquad i > g. \tag{4-9}$$

Computation Note: *It's very easy to compare alternatives with a perpetual service time because the annual worth limits don't require factor values!*

EXAMPLE 4.7 Finding the Least-Cost Bridge: An Annual Worth Comparison with a Perpetual Service Time (*AWC*-Class)

Problem: Table 4.16 lists the economic data for three proposed bridges. *If the MARR% is 7% per year and the service time is perpetual, which bridge is the least-cost alternative?*

TABLE 4.16 Economic Data for Bridges *X*, *Y*, and *Z*

DATA TYPE	*X*	*Y*	*Z*
Purchase Price: P	$\$60.0 \times 10^6$	$\$65.0 \times 10^6$	$\$67.0 \times 10^6$
First Year Operating Cost: A	$\$1.75 \times 10^6$	$\$1.50 \times 10^6$	$\$1.00 \times 10^6$
Annual Operating Cost Increment: G	$\$0.100 \times 10^6$	$\$0.075 \times 10^6$	$\$0.050 \times 10^6$
First Year Labor Cost: A_1	$\$1.25 \times 10^6$	$\$1.25 \times 10^6$	$\$1.35 \times 10^6$
Annual Rate of Change of Labor Cost: $g\%$	3%	3%	3%

Solution: Table 4.17 on the following page compares the annual worths of the cash flow elements of the three bridges when the MARR% is 7% per year. As you can see, *the Z bridge is the least cost alternative.*

Computation Note: *All worths in Table 4.17 are rounded to three significant digits because all numbers magnitudes are specified with this number of significant digits.*

TABLE 4.17 Annual Perpetual Worth Comparison of Bridges X, Y, and Z: $ST = ST_\infty$ and $r^*\% = 7\%$ per Year

WORTH TYPE	BRIDGE ANNUAL WORTH TABLE		
	X	Y	Z
$AW_{P,\infty}$	$Pr^* = \$60.0 \times 10^6 \times 0.07 = \$4.20 \times 10^6$	$Pr^* = \$65.0 \times 10^6 \times 0.07 = \$4.55 \times 10^6$	$Pr^* = \$67.0 \times 10^6 \times 0.07 = \$4.69 \times 10^6$
$AW_{A,\infty}$	$\$1.75 \times 10^6$	$\$1.50 \times 10^6$	$\$1.00 \times 10^6$
$AW_{G,\infty}$	$\frac{G}{r^*} = \frac{\$0.100 \times 10^6}{0.07} = \$1.43 \times 10^6$	$\frac{G}{r^*} = \frac{\$0.075 \times 10^6}{0.07} = \$1.07 \times 10^6$	$\frac{G}{r^*} = \frac{\$0.050 \times 10^6}{0.07} = \$0.714 \times 10^6$
$AW_{A_1,\infty}$	$A_1\frac{r^*}{r^* - g} = \$1.25 \times 10^6 \frac{0.07}{0.04} = \$2.19 \times 10^6$	$A_1\frac{r^*}{r^* - g} = \$1.25 \times 10^6 \frac{0.07}{0.04} = \$2.19 \times 10^6$	$A_1\frac{r^*}{r^* - g} = \$1.35 \times 10^6 \frac{0.07}{0.04} = \$2.36 \times 10^6$
ΣAW_∞	**$\$9.57 \times 10^6$**	**$\$9.31 \times 10^6$**	**$\$8.76 \times 10^6$**

4.6 ANNUAL WORTH INVESTMENT CALCULATIONS

The General Repayment Equation (GRE) is the starting point for calculating a simple investment's IRR:

$$\sum_{k=0}^{N} \frac{R_k}{(1+i)^k} = \sum_{k=0}^{N} \frac{P_k}{(1+i)^k}.$$

However, ***we can specify a value for the IRR and use it to calculate the value of any cash flow in the GRE.***

Equation (2-37) shows that the present worths of the receipts and disbursements of a *simple* investment are equal. Therefore, the receipt and disbursement series are equivalent. But this also means that ***the annual worths of the receipts and disbursements on the investment's life are equal.*** Hence, we can make a variety of annual worth ***equivalence*** calculations for an investment with a ***specified* IRR**. In all of them we solve for *one* unknown. The following list describes three common types of equivalence calculations that we can make for a *specified* IRR:

1. Calculate the metric of a receipt series that yields a specified IRR%.
2. Calculate the metric of a disbursement series that yields a specified IRR%.
3. Calculate the counteroffer for an investment that yields a specified IRR%.

EXAMPLE 4.8 Calculating the Required Annual Receipt of an Investment with an Annual Worth Calculation (*E*-Class)

Problem: The equipment to produce a small automotive part costs \$125,000 (This is the *specified* value of the investment *P*.) and has a salvage value *S* of \$6,000 at the end of 10 years. The *annual* disbursements A_D are \$3,000 per year. *What annual receipt A_R from the sales of the automotive part is required to yield an IRR% of 12% per year?*

Solution: As with any simple investment, the annual worths of the disbursements and receipts on the investment's life are equal. Hence,

$$\underbrace{A_R + S(A/F, 12\%, 10)}_{AW \text{ of Receipts}} = \underbrace{P(A/P, 12\%, 10) + A_D}_{AW \text{ of Disbursements}}.$$

Hence,

$$A_R = \underbrace{P}_{\$125{,}000} \underbrace{(A/P, 12\%, 10)}_{0.17698} + \underbrace{A_D}_{\$3{,}000} - \underbrace{S}_{\$6{,}000} \underbrace{(A/F, 12\%, 10)}_{0.056984}$$
$$= \$22{,}123 + \$3{,}000 - \$342 = \$24{,}781.$$

The following spreadsheet further illustrates that ***we can solve for any unknown in the General Repayment Equation***:

	A	B	C	D
1				
2		***IRR% of the Automotive Parts Investment***		
3		***k***	**Cash Flow, \$**	
4		0	(125,000)	P
5		1	21,781	
6		2	21,781	
7		3	21,781	
8		4	21,781	
9		5	21,781	$A_R - A_D$
10		6	21,781	
11		7	21,781	
12		8	21,781	
13		9	21,781	
14		10	27,781	$A_R - A_D + S$
15			**12.00%**	**=IRR(C4:C14)**

Spreadsheet 4.4 NOTE: We originally found the value of A_R when the value of IRR% was specified as 12%. Here we specify the value of A_R and find the value of IRR%.

EXAMPLE 4.9 Calculating a Counteroffer for a Pizza Parlor Investment with an Annual Worth Calculation (*E*-Class)

Problem: The owner of a pizza parlor wants to sell it for \$500,000 so that he can retire to Florida. His accountant certifies that the annual receipts A_R for the past two years were \$350,000 per year. The annual operating cost A_D for ingredients, beverages, utilities, maintenance, and labor is expected to be \$250,000 this year, but it has been increasing at a rate of 3% per year. Your uncle is interested in purchasing the pizza parlor, but he is concerned that the annual receipts are stagnant while the annual operating costs are rising. Hence, he would like to earn 20% per year on the investment over the next nine years to account for this risk. *What should his counteroffer P be if the market value of the pizza parlor in nine years is assumed to be P?*

Solution: As with any simple investment, the annual worths of the disbursements and receipts on the investment's life, which is nine years, are equal:

$$\underbrace{A_R + MV(A/F, 20\%, 9)}_{AW\text{ of Receipts}} = \underbrace{P(A/P, 20\%, 9) + A_D(A/A_1, 3\%, 20\%, 9)}_{AW\text{ of Disbursements}}.$$

Hence,

$$\underbrace{P(A/P, 20\%, 9)}_{0.24808} - \underbrace{P(A/F, 20\%, 9)}_{0.048079} = \underbrace{A_R}_{\$350{,}000} - \underbrace{A_D}_{\$250{,}000}\underbrace{(A/A_1, 3\%, 20\%, 9)}_{1.09028},$$

and the value of P rounded to three significant digits is given by

$$P = \frac{\$350{,}000 - \$250{,}000 \times 1.09028}{0.24808 - 0.048079} = \underline{\underline{\$387{,}000.}}$$

This is \$113,000 less than the owner's asking price. However, your uncle should present the \$387,000 offer. *He will never know if it will be accepted unless he does so.* Meanwhile, he should try to figure out why the receipts have been flat. Is new competition the problem, or is the existing competition more creative in its pizza offerings? Maybe its a combination of both of these reasons. Does your uncle have any ideas that will increase the receipts, decrease costs, or both? Lastly, a value of 20% per year for the IRR% is an excellent rate of return for a relatively low-risk investment. Your uncle might eventually have to settle for a *lower* value of the IRR% and a *larger* value of *P*.

4.7 SUMMARY

Section 4.1 defines the quantities used in an annual worth least-cost study: cost alternatives, minimum attractive rate of return (MARR), specified service time *ST*, Annual Worth Comparison Rule, and the Consistency Equation.

Subsection 4.2.1 explains how to use an annual worth table to make a least-cost study of *two or more* cost alternatives with a *specified* service time. The ***ANNUAL WORTH COMPARISON Calculator*** is used in Subsection 4.2.2 to automate the calculations. Subsection 4.2.4 shows how to incorporate the *replacement* cost(s) of a component of each cost alternative into a least-cost study. The ***ANNUAL WORTH COMPARISON Calculator with Component Replacement*** is used in Subsection 4.2.5 to automate the calculations. Lastly, it is shown in Subsection 4.2.6 that *the alternative with the smallest purchase price always becomes the least-cost alternative as the value of the MARR increases.*

Subsection 4.3.1 defines the least common multiple *LCM* service time for *two* alternatives with an *indefinite* service time. Subsection 4.3.2 show how to use the **Repeatability Assumption** (RA) to make an annual worth comparison of *two* cost alternatives with an *LCM* service time. Subsection 4.3.3 compares two reactors (previously compared with a specified service time) with an *LCM* service time and the RA. The disadvantages of using an *LCM* service time are explained in Subsection 4.3.4. To eliminate them, a specified service time is used in subsequent least-cost studies in the book.

Section 4.4 explains three other types of cost alternatives: **extra capacity**, **dissimilar**, and **irregular**. Examples 4.4, 4.5, and 4.6 show how to compare their annual worths. Section 4.5 explains how to compare two or more cost alternatives in ***perpetual*** service and illustrates the procedure with Example 4.7. Lastly, Section 4.6 shows that we can make various types of annual worth *equivalence* calculations for investments with a *specified* IRR. In all of them we solve for *one* unknown. Examples 4.8 and 4.9 illustrate these types of calculations.

4.8 PROBLEMS

Computation Note: *If you can't use one of the* ***ANNUAL WORTH COMPARISON Calculators*** *to solve a problem, don't bother with a worth table to present your calculations. I use worth tables because they clearly contrast the worth difference between corresponding cash flow elements of the alternatives. However, it's easier to calculate the worths of the elements of each alternative on separate lines. NOTE: When there are three or more alternatives, a worth table is impractical. Here are the separate line worth calculations of compressor X in Worth Table 4.2:*

Compressor X

$$AW_P = \$80{,}000\underbrace{(A/P, 20\%, 10)}_{0.23852} = \$19{,}082$$

$$AW_S = \$4{,}000\underbrace{(A/F, 20\%, 10)}_{0.03852} = (\$154)$$

$$AW_A = \$3{,}000$$

$$NAW_{A_1} = \$750\left[\begin{array}{c}\underbrace{(A/A_1, 5\%, 20\%, 10)}_{1.17182} + \\ -\underbrace{(P/F, 20\%, 1)}_{0.83333}\underbrace{(A/P, 20\%, 10)}_{0.23852}\end{array}\right] = \$730$$

$$\Sigma AW = \underline{\underline{\$22{,}658}}$$

(P4-1) Here is the economic data for marble quarry dump trucks *X*, *Y*, and *Z*. *Which truck is the least-cost alternative when the MARR% is 11% per year and the anticipated service time is seven years?*

TABLE 4.18 Economic Data for the Dump Trucks

DATA TYPE	*X*	Y	Z
Truck Purchase Price: *P*	$\$1.57 \times 10^6$	$\$1.6 \times 10^6$	$\$1.8 \times 10^6$
Salvage Value: *S*	\$75,000	\$80,000	\$100,000
Initial Transmission Cost: C_0	\$90,000	\$100,000	\$120,000
Transmission Life: *l*	3a	4a	5 a

TABLE 4.18 Economic Data for the Dump Trucks *(continued)*

Annual Rate of Change of Transmission Cost: $f\%$	4%	4%	4%
First Year Fuel Cost: A	\$33,500	\$32,000	\$30,000
Annual Fuel Cost Increment: G	\$1,675	\$1,600	\$1,500
First Year Maintenance Cost[1]: A_1	\$3,000	\$3,200	\$3,300
Annual Rate of Change of Maintenance Cost: $g\%$	3.5%	3.5%	3.5%
Service Time: ST		7a	

[1] The first year cost is covered by the manufacturer's warranty.

(P4-2) A company is currently using manual labor to remove defective parts from the conveyor belt that transports the parts to packaging for shipment to the company's customers. The current annual labor cost for removing defective parts is \$145,000, but it has been increasing at the rate of 5% per year. The *Detecto* and *Removo* machines automatically detect and remove the defective parts; their economic data is listed in Table 4.19. *What should the company do for the next 10 years if the MARR% is 12.5% per year?*

TABLE 4.19 Economic Data for the Parts Removal Equipment

DATA TYPE	*Detecto*	*Removo*
Machine Purchase Price: P	\$350,000	\$425,000
Salvage Value: S	\$18,000	\$25,000
Initial Mechanism Cost: C_0	\$65,000	\$80,000
Mechanism Life: l	4a	5a
Annual Rate of Change of Mechanism Cost: $f\%$	5%	5%
First Year Electricity Cost: A	\$3,500	\$4,000
Annual Increase of Electricity Cost: G	\$350	\$400
First Year Maintenance Cost[1]: A_1	\$1,350	\$1,475
Annual Rate of Change of Maintenance Costs: $g\%$	6%	6%
Service Time: ST	10a	

[1] The first year cost is covered by the manufacturer's warranty.

(P4-3) After Hurricane Sandy a homeowner wanted to install an emergency electric power generator fueled by propane from two portable tanks. Table 4.20 lists the economic data provided by three reputable local contractors C_1, C_2, and C_3 for the equipment installed on a pressure treated wood platform supported on concrete footings. *Which proposal would you have recommended if the MARR% was 8% per year?*

TABLE 4.20 Economic Data for Propane-Fueled Electric Power Generators

DATA TYPE	C_1	C_2	C_3
Installed Cost: P	**\$21,000**	**\$23,000**	**\$24,500**
Salvage Value: S	\$1,500	\$1,700	\$1,950

TABLE 4.20 Economic Data for Propane-Fueled Electric Power Generators *(continued)*

DATA TYPE	C_1	C_2	C_3
Annual Maintenance Cost: A	$800	$750	$700
Annual Maintenance Cost Increment: G	**$150**	**$145**	**$130**
First Year Propane Cost[1]: A_1	$850	$810	$795
Annual Rate of Change of Propane Cost: $g\%$	4%	4%	4%
Service Time: *ST*		**10a**	

[1] Historical utility records indicate that power is lost two to three times a year for a total outage of 14 days.

(P4-4) A small town's potable water supply system (PWSS) is badly corroded and must be replaced at a cost of $950,000. However, a member of the Town Board, which is responsible for authorizing the project, notes that the town's population has recently started to increase. A consultant reports that in eight years the proposed $950,000 PWSS will not have the capacity to handle the increased demand. This will require an *additional* PWSS in parallel with it that will cost $1,200,000. She recommends that a *larger* PWSS that costs $1,900,000 should be built *now* to meet the present and future demands. Table 4.21 summarizes the consultant's data. *Which PWSS would you recommend if the MARR% is 7% per year?*

TABLE 4.21 Data for the Potable Water Supply Systems

DATA TYPE	X	Y
PWSS Purchase Price for X_{1st} and Y: P	$950,000	$1,900,000
PWSS Purchase Price for X_{2nd} at TP_8: P_{2nd}	$1,200,000	—
First Year Operating Cost for X_{1st} and Y: A_1	$50,000	$75,000
Annual Rate of Change of Operating Cost for X_{1st} and Y: $g\%$	5%	5%
First Year Operating Cost for X_{2nd}: $A_{1,2nd}$	$80,000	—
Annual Rate of Change of Operating Cost for X_{2nd}: $g_{2nd}\%$	5%	—
Service Time: *ST*	50 a	

(P4-5) *If the receipt series in Example 4.8 is expected to be a geometric gradient series rather than a uniform series, what is the first receipt A_1 if the value of $g\%$ is 4%?* Nothing else changes.

5

Present Worth Comparisons

OBJECTIVES

Subsection 4.1.1 provides the *identical* definitions used in annual *and* present worth comparisons, and Subsection 4.1.2 shows that *annual worth and present worth comparisons always identify the same least-cost alternative* because the worths of the cash flows of *any* alternative[1] X are related by the Consistency Equation (4-1):

$$\boxed{AW_X = PW_X(A/P, r^*, ST).} \tag{4-1}$$

After reading this chapter you will be able to perform *present worth comparison* studies, ***PWC*-Class** problems, to

1. Find the least-cost alternative of two or more alternatives with the same or different lives, component replacements, and a specified service time.
2. Find the least-cost alternative for three special classes of cost alternatives: Extra Capacity, Dissimilar, and Irregular.
3. Find the least-cost alternative when two or more cost alternatives provide perpetual service.
4. Make present worth investment equivalence calculations.

In addition, *you will be able to choose between the annual and present worth comparison methods for a particular worth comparison.*

To conveniently compare the two worth comparison methods, we just solve some of the examples in Chapter 4 with present worth calculations.

5.1 SPECIFIED SERVICE TIME

5.1.1 Present Worths Without Replacement Costs

EXAMPLE 5.1 Finding the Least-Cost Compressor: A Present Worth Comparison When $ST = L = 10a$ (*PWC*-Class)

Problem: Table 5.1 lists the economic data used for compressors X and Y in Example 4.1. The annual maintenance costs are represented by a geometric gradient series, and the MARR% is 20% per year.

[1]As in Chapter 4, "alternative" means "cost alternative."

TABLE 5.1 Economic Data for the Compressors

DATA TYPE	*X*	*Y*
Purchase Price: *P*	\$80,000	\$85,000
Salvage Value: *S*	\$4,000	\$4,500
Annual Energy Cost: *A*	\$3,000	\$2,200
First Year Maintenance Cost: A_1	\$750	\$460
Annual Rate of Change of Maintenance Costs: *g*%	5%	5%
Life: *L*	10 a	10 a

(1) Use a present worth comparison to find the least-cost compressor when the maintenance cost of each compressor in the first year is covered by its manufacturer's warranty. (2) Show that the results are consistent with those of Example 4.1.

Solution: ***Find the Least-Cost Compressor***

Table 5.2 shows the **present worth table** used to present the calculations. It clearly displays and contrasts the present worths of each compressor's cash flow elements. *As in the uniform worth comparison in Example 4.1, compressor X is the least-cost alternative.*

TABLE 5.2 Comparing the Costs of Compressors *X* and *Y*: $ST = L = 10$ a; $r^{*}\% = 20\%$ per Year[1]

WORTH TYPE	COMPRESSOR PRESENT WORTH TABLE	
	X	*Y*
PW_P	\$80,000	\$85,000
PW_S	$(\$4{,}000)\underbrace{(P/F, 20\%, 10)}_{0.16151} = (\$646)$	$(\$4{,}500)\underbrace{(P/F, 20\%, 10)}_{0.16151} = (\$727)$
PW_A	$\$3{,}000\underbrace{(P/A, 20\%, 10)}_{4.19247} = \$12{,}577$	$\$2{,}200\underbrace{(P/A, 20\%, 10)}_{4.19247} = \$9{,}223$
NPW_{A_1}	$\$750\left[\underbrace{(P/A_1, 5\%, 20\%, 10)}_{4.91283} + -\underbrace{(P/F, 20\%, 1)}_{0.83333}\right] = \$3{,}060$	$\$460\left[\underbrace{(P/A_1, 5\%, 20\%, 10)}_{4.91283} + -\underbrace{(P/F, 20\%, 1)}_{0.83333}\right] = \$1{,}877$
ΣPW	**\$94,991**	**\$95,373**

[1] Worth entries in cost comparison tables or spreadsheets are always rounded to the dollar because of the uncertainty in predicting their values over many interest periods.

We now use Equation (4-1) to show that *the results are consistent with those of Example 4.1*:

$$AW_X = \underbrace{PW_X}_{\$94{,}991}\underbrace{(A/P, 20\%, 10)}_{0.23852} = \underline{\underline{\$22{,}657}}$$

$$AW_Y = \underbrace{PW_Y}_{\$95{,}373}\underbrace{(A/P, 20\%, 10)}_{0.23852} = \underline{\underline{\$22{,}748.}}$$

5.1.2 *PRESENT WORTH COMPARISON Calculator*

Spreadsheet 5.1 is the ***PRESENT WORTH COMPARISON Calculator***, an Excel spreadsheet created by the author and located in the *Book Templates* file. We use it to perform a present worth comparison of two or more cost alternatives with lives that equal or *exceed* the service time. The cash flow elements consist of a purchase price, a salvage or **market value**,[2] and *any* combination of the three basic series. *The spreadsheet automatically calculates the values of the present worths and their sums, for arbitrary values of the INPUT variables when g% and i% are different.*

	A	B	C	D	E	F	G	H
1								
2		***PRESENT WORTH COMPARISON Calculator***						
3			***Input:***			***Solution: Present Worths and Sums***		
4			*W*	*X*	*Y*	*W*	*X*	*Y*
5		***P***	$75,000	$80,000	$85,000	$75,000	$80,000	$85,000
6		***S ⊕ MV***	($3,500)	($4,000)	($4,500)	($565)	($646)	($727)
7		***A***	$3,350	$3,000	$2,200	$14,045	$12,577	$9,223
8		***G***	$950	$0	$0	$12,243	$0	$0
9		***A_1***	$0	$750	$460	$0	$3,060	$1,877
10		***g%***	0%	5%	5%			
11		***$A_{1,net}$***	0	1	1			
12		***r*%***	20%		***ΣPW***	**$100,723**	**$94,991**	**$95,373**
13		***ST***	10					

Spreadsheet 5.1 ***PRESENT WORTH COMPARISON Calculator*** NOTES: (1) Alternatives *X* and *Y* have the *same* economic data and present worths as compressors *X* and *Y* in Example 5.1. (2) The symbol ***S ⊕ MV*** in cell B6 means ***salvage value*** *S or* ***market value*** *MV*. (3) Placing a thick black border on the present worth cell of the least-cost alternative quickly identifies it.

5.1.3 Using the *PRESENT WORTH COMPARISON Calculator*

To use the ***PRESENT WORTH COMPARISON Calculator*** in a least-cost study, make a copy of it and enter the values of the variables in the *INPUT* columns. NOTE: As soon as you change the value of an alternative's input variable, the spreadsheet automatically changes the values of the corresponding present worth and total present worth in the *SOLUTION* cells.

Specifying a Present Worth or Net Present Worth

When a value in array C11:E11 is 0, the present worth of the geometric gradient series is calculated. When a value is 1, the net present worth of the geometric gradient series is

[2] *When the service time is less than the life of an alternative, we use its market value in a least-cost study.* The **market value** *MV* of a piece of equipment, such as a compressor or a heat exchanger, is its value in the used equipment market. It is generally greater than the equipment's salvage value.

calculated. For example, the net present worths of the geometric gradient series of alternatives X and Y are calculated in Spreadsheet 5.1.

Adding More Cost Alternatives

If you want to *add* more cost alternatives to the ***PRESENT WORTH COMPARISON Calculator***, make a copy of it and do the following:

1. *Select* one or more *adjacent* columns to the *right* of the E column > *Right Click* the selected column(s) > *Click* **Insert**.
2. *Enter* the name(s) of the cost alternative(s) in Row 4 of the inserted *INPUT* column(s).
3. *Enter* the values of P, $S \oplus MV$, A, G, A_1, $g\%$, and $A_{1,\text{net}}$ in Rows 5 through 11 of the inserted *INPUT* column(s).
4. *Paste* the *original* array H5:H13 into the *empty SOLUTION* column(s) to the *right* of the *original* H column to evaluate the solution cells of the added cost alternative(s) > *Enter* the name(s) in Row 4 of the *SOLUTION* column(s).
5. *Merge* the cells in each title row.

5.1.4 Component Replacement Costs

As explained Subsection 4.2.4, an alternative may have a component that must be replaced during its life. Hence, the **component replacement cost** at the end of the kth year C_k is defined as

$$C_k \equiv C_0(1+f)^k, \tag{5-1}$$

where C_0 is the cost at the *beginning* of the service time, and f is the annual rate at which it increases. It depends on the prevailing economic climate, and a range of $0.02 < f < 0.05$ is typical. Example 5.2 illustrates the use of Equation (5-1).

EXAMPLE 5.2 Finding the Least-Cost Reactor: A Present Worth Comparison with Component Replacement Costs When $L > ST = 10$ a (*PWC*-Class)

Problem: A chemical company is considering the construction of a plant to produce cumene, which is a paint thinner and a component of high octane fuels. Two American process engineering companies have submitted bids to supply their proprietary catalytic reactors to carry out the reaction. As with the compressors for a chemical plant, the MARR% is 20% per year and the service time is 10 years.

The cost of reactor X *and* its catalyst is \$165,000, and the *estimated* market value of the reactor at the end of the service time is \$11,800. The initial cost of the catalyst is \$34,000. It deteriorates with time and must be replaced every four years. The energy cost is \$4,000 in the first year, and it is *expected* to increase by \$250 in each subsequent year of the service time.

The cost of Reactor Z *and* its catalyst is \$200,000, and the *estimated* market value of the reactor at the end of the service time is \$25,400. The initial cost of the catalyst is \$22,000. It too deteriorates with time and must be replaced every five years. The energy cost is \$1,200 in the first year, and it is *expected* to increase by \$150 in each subsequent year in the service time. *(1) Use a present worth comparison to find the least-cost reactor. (2) Show that the results are consistent with those of Example 4.2.* NOTE: All replacement costs increase at a rate of 3% per year.

Solution: Table 5.3 lists the economic data for reactors X and Z.

TABLE 5.3 Economic Data for Reactors *X* and *Z*

DATA TYPE	*X*	*Z*
Reactor Purchase Price: *P*	\$165,000	\$200,000
Reactor Life: *L*	12 a	15 a
Market Value: *MV*	\$11,800	\$25,400
Initial Catalyst Cost: C_0	\$34,000	\$22,000
Annual Percentage Increase of Catalyst Cost: *f*%	3%	3%
Catalyst Life: *l*	4 a	5 a
First Year Energy Cost: *A*	\$4,000	\$1,200
Annual Energy Cost Increment: *G*	\$250	\$150
Service Time: *ST*	10 a	

Book Convention: *Tabulating economic data that is presented in a narrative form clearly reveals the differences between the alternatives and minimizes the chance that one or more incorrect data values will be used in a least-cost study. This convention is followed throughout the book.*

Table 5.4 is a present worth table that displays the present worths of the cash flow elements of reactors *X* and *Z* when the MARR% is 20% per year.

TABLE 5.4 Comparing the Costs of Reactors *X* and *Z*: $L > ST = 10$ a and $r^{*}\% = 20\%$ per Year

WORTH TYPE	REACTOR PRESENT WORTH TABLE	
	X	*Z*
PW_P	\$165,000	\$200,000
PW_{MV}	$(\$11{,}800)\underbrace{(P/F,20\%,10)}_{0.16151} = (\$1{,}906)$	$(\$25{,}400)\underbrace{(P/F,20\%,10)}_{0.16151} = (\$4{,}102)$
PW_{C_k}	$\$34{,}000\left[\begin{array}{l}(1+0.03)^4\underbrace{(P/F,20\%,4)}_{0.48225}+\\ +(1+0.03)^8\underbrace{(P/F,20\%,8)}_{0.23257}\end{array}\right] = \$28{,}471$	$\$22{,}000(1+0.03)^5\underbrace{(P/F,20\%,5)}_{0.40188} = \$10{,}250$
PW_A	$\$4{,}000\underbrace{(P/A,20\%,10)}_{4.19247} = \$16{,}770$	$\$1{,}200\underbrace{(P/A,20\%,10)}_{4.19247} = \$5{,}031$
PW_G	$\$250\underbrace{(P/G,20\%,10)}_{12.88708} = \$3{,}222$	$\$150\underbrace{(P/G,20\%,10)}_{12.28708} = \$1{,}933$
ΣPW	**\$211,557**	\$213,112

Reactor X is the least-cost reactor.

Consistency Check

We now use Equation (4-1) to show that *the results are consistent with those of Example 4.2*:

$$AW_X = \underbrace{PW_X}_{\$211,557} \underbrace{(A/P, 20\%, 10)}_{0.23852} = \underline{\underline{\$50,461}}$$

$$AW_Z = \underbrace{PW_Z}_{\$213,112} \underbrace{(A/P, 20\%, 10)}_{0.23852} = \underline{\underline{\$50,831}}.$$

5.1.5 Adding Component Replacement Costs to the *PRESENT WORTH COMPARISON Calculator*

Spreadsheet 5.3 shows the ***PRESENT WORTH COMPARISON Calculator with COMPONENT REPLACEMENT*** spreadsheet in the *Book Templates* file. It is the ***PRESENT WORTH COMPARISON Calculator*** with the addition of *new* rows 7, 8, 9, and 10 for the initial cost of a component C_0, its life l (measured in years), and the annual rate of change of its cost f. The symbol $\#_{max} = 2$ specifies the maximum number of component replacements in this spreadsheet: *one* replacement (value 0) or *two* replacements (value 1) in the *ST* interest periods. *If an alternative has no replacement costs, enter zero for C_0, l, f, and $\#_{max}$.*

	A	B	C	D	E	F	G	H
1								
2		***PRESENT WORTH COMPARISON Calculator with Component Replacement***						
3		***INPUT***				***SOLUTION Present Worths and Sums***		
4			***X***	***Y***	***Z***	***X***	***Y***	***Z***
5		***P***	$165,000	$185,000	$200,000	$165,000	$185,000	$200,000
6		***S ⊕ MV***	($11,800)	($23,000)	($25,400)	($1,906)	($3,715)	($4,102)
7		C_0	$34,000	$28,000	$22,000	$28,471	$13,690	$10,249
8		l	4	5	5			
9		f	3%	4%	3%			
10		$\#_{max} = 2$	1	0	0			
11		***A***	$4,000	$1,800	$1,200	$16,770	$7,546	$5,031
12		***G***	$250	$175	$150	$3,222	$2,255	$1,933
13		A_1	$0	$300	$0	$0	$1,177	$0
14		***g%***	0%	4%	0%			
15		$A_{1,net}$	0	1	0			
16		***r*%***	20.00%		**ΣPW**	**$211,557**	**$205,953**	**$213,111**
17		***ST***	10					

Spreadsheet 5.2 ***PRESENT WORTH COMPARISON Calculator with Component Replacement(s)*** NOTE: This spreadsheet calculates the present worths of reactors *X* and *Z* in Example 5.2 and the least-cost reactor *Y*.

5.2 OTHER CAPACITY ALTERNATIVES

5.2.1 Extra Capacity & Staged Construction Alternatives

Subsection 4.4.1 defines an **extra capacity** alternative as one that satisfies a present and future demand and **staged construction** as two smaller sequential alternatives that also satisfy the present and future demands. Example 4.4 compares the annual worths of the these types of alternatives. Example 5.3 compares the present worths of the *same* alternatives.

EXAMPLE 5.3 Finding the Least-Cost Elementary School When One Is an Extra Capacity Alternative: A Present Worth Comparison with a Specified Service Time and Staged Construction (*PWC*-Class)

Problem: Table 5.5 lists the economic data for the two alternatives.

TABLE 5.5 Economic Data for Elementary Schools *X* (Two Schools) and *Y* (One School)

DATA TYPE	X	Y
School Price for X_{1st} and Y: P	$\$1.2 \times 10^6$	$\$2 \times 10^6$
School Price for X_{2nd} at TP_6: $P_{2nd} = \$1.2 \times 10^6 \times (1.03)^6$	$\$1.433 \times 10^6$	—
First Year Operating Cost for X_{1st} and Y: A	\$130,000	\$180,000
Annual Operating Cost Increment for X_{1st} and Y: G	\$3,750	\$6,000
First Year Operating Cost for X_{2nd} at TP_7: A_{2nd}	\$152,500	—
Annual Operating Cost Increment for X_{2nd} at TP_8: G_{2nd}	\$3,750	—

(1) Use a present worth comparison to find the least-cost alternative when the MARR% is 8% per year,[3] $ST = 15$ a, *and the market values of the buildings can be neglected. (2) Show that the results are consistent with those of Example 4.4.*

Solution: ***Find the Least-Cost School Alternative***

Table 5.6 compares the present worths of the cash flow elements of the *X* and *Y* alternatives. As you can see, *the extra capacity school Y is the least-cost alternative.*

TABLE 5.6 Present Worth Comparison of Alternatives *X* (Two Schools) and *Y* (One School): $ST = 15$ a and $r^*\% = 8\%$ per Year

WORTH TYPE	ELEMENTARY SCHOOL PRESENT WORTH TABLE	
	X	Y
PW_P	$\$1.2 \times 10^6$	$\$2 \times 10^6$
$PW_{P_{2nd}}$	$\$1.433 \times 10^6 \underbrace{(P/F, 8\%, 6)}_{0.63017} = \$903{,}034$	—
PW_A	$\$130{,}000 \underbrace{(P/A, 8\%, 15)}_{8.55948} = \$1,112,732$	$\$180{,}000 \underbrace{(P/A, 8\%, 15)}_{8.55948} = \$1,540,706$

[3] The small MARR% represents the interest rate of the bonds that will be issued to finance the least-cost alternative.

TABLE 5.6 Present Worth Comparison of Alternatives X (Two Schools) and Y (One School): $ST = 15$ a and $r^*\% = 8\%$ per Year *(continued)*

Worth Type	Elementary School Present Worth Table	
	X	Y
PW_G	$\$3{,}750\underbrace{(P/G, 8\%, 15)}_{47.88566} = \$179{,}571$	$\$6{,}000\underbrace{(P/G, 8\%, 15)}_{47.88566} = \$287{,}314$
$PW_{A_{2nd}}$	$\$152{,}500\underbrace{(F/A, 8\%, 9)}_{12.48756}\underbrace{(P/F, 8\%, 15)}_{0.31524} = \$600{,}328$	—
$PW_{G_{2nd}}$	$\$3{,}750\underbrace{(F/G, 8\%, 9)}_{43.59447}\underbrace{(P/F, 8\%, 15)}_{0.31524} = \$51{,}535$	—
ΣPW	**$4,047,200**	**$3,828,020**

Consistency Check

We now use Equation (4-1) to show that *the results are consistent with those of Example 4.4*:

$$AW_X = \underbrace{PW_X}_{\$4{,}047{,}200}\underbrace{(A/P, 8\%, 15)}_{0.11683} = \underline{\underline{\$472{,}834}}$$

$$AW_Y = \underbrace{PW_Y}_{\$3{,}828{,}020}\underbrace{(A/P, 8\%, 15)}_{0.11683} = \underline{\underline{\$447{,}228}}.$$

5.2.2 Dissimilar & Irregular Alternatives

Dissimilar alternatives are compared just like we compare similar alternatives. However, calculating the present worths of **irregular** alternatives is simpler than calculating their annual worths because the first step in doing so is the calculation of their present worths. See Step-by-Step 4.1 in Subsection 4.4.3.

5.3 PERPETUAL SERVICE TIME

5.3.1 Definitions

As stated in Chapter 4, our national infrastructure contains structures that will essentially last "forever" if they are properly maintained. That is, they are designed for a **perpetual service time** ST_∞. To make a present worth comparison of alternatives designed for a perpetual service time, we need some or all of the present worth limits of an annual series, gradient series, and geometric gradient series as the number of interest periods approaches infinity. These present worth limits are called **capitalized costs** CC, and their simple derivations are shown on the following page.

5.3.2 Capitalized Costs of the Basic Series

Annual Series

The capitalized cost of an annual series is obtained from Equation (2-46):

$$\boxed{CC_A \equiv \lim_{N\to\infty} PW_A = A\lim_{N\to\infty}\frac{1-(1+i)^{-N}}{i} = \frac{A}{i}.} \tag{5-2}$$

Gradient Series

The definition of the capitalized cost of a gradient series (see Equation (2-81)) yields an indeterminate result:

$$CC_G \equiv \lim_{N\to\infty} PW_G = G\lim_{N\to\infty}\left[\frac{(1+i)^N - Ni - 1}{i^2(1+i)^N}\right] = G\lim_{N\to\infty}\left[\frac{1}{i^2} - \frac{N}{i(1+i)^N} - \frac{1}{i^2(1+i)^N}\right] =$$
$$= G\left[\frac{1}{i^2} - \frac{\infty}{i(1+i)^\infty}\right].$$

Hence, we apply L'Hopital's Rule to find the limit:

$$CC_G = \frac{G}{i^2} - G\lim_{N\to\infty}\left[i^{-1}\frac{\frac{d}{dN}N}{\frac{d}{dN}(1+i)^N}\right] = \frac{G}{i^2} - \underbrace{G\lim_{N\to\infty}\left[i^{-1}\frac{1}{N(1+i)^{N-1}}\right]}_{0}.$$

Thus,

$$\boxed{CC_G = \frac{G}{i^2}.} \tag{5-3}$$

Geometric Gradient Series

The definition of the capitalized cost of a geometric gradient series (see Equation (2-95)) is

$$CC_{A_1} \equiv \lim_{N\to\infty} PW_{A_1} = A_1\lim_{N\to\infty}\left[\frac{1-(1+g)^N(1+i)^{-N}}{i-g}\right]. \tag{5-4}$$

When $i > g$, the product in the numerator converges to 0. Hence,

$$\boxed{CC_{A_1} = \frac{A_1}{i-g} \qquad i > g.} \tag{5-5}$$

5.3.3 Comparison of Annual Worth Limits & Capitalized Costs

Table 5.7 lists the four annual worth limits and capitalized costs.

TABLE 5.7 Comparison of Annual Worth Limits and Capitalized Costs

COST VARIABLE	ANNUAL WORTH LIMIT	CAPITALIZED COST
P	Pi	P
A	A	$\frac{A}{i}$
G	$\frac{G}{i}$	$\frac{G}{i^2}$
A_1	$A_1 \frac{i}{i-g}$	$\frac{A_1}{i-g}$

Computation Notes:*(1) An annual worth limit is the product of the corresponding capitalized cost and the interest rate.*
(2) It's very easy to compare alternatives with a perpetual service time because the annual worth limits and capitalized costs don't require factor values!

EXAMPLE 5.4 Finding the Least-Cost Bridge: A Capitalized Cost Comparison (*PWC*-Class)

Problem: Table 5.8 lists the economic data for the three proposed bridges in Example 4.7. *(1) Use a capitalized cost comparison to find the least-cost bridge when the MARR% is 7% per year and the service time is perpetual. (2) Show that the results are consistent with those of Example 4.7.*

TABLE 5.8 Economic Data for Bridges *X*, *Y*, and *Z*

DATA TYPE	X	Y	Z
Purchase Price: P	$\$60.0 \times 10^6$	$\$65.0 \times 10^6$	$\$67.0 \times 10^6$
First Year Operating Cost: A	$\$1.75 \times 10^6$	$\$1.50 \times 10^6$	$\$1.00 \times 10^6$
Annual Operating Cost Increment: G	$\$0.100 \times 10^6$	$\$0.075 \times 10^6$	$\$0.050 \times 10^6$
First Year Labor Cost: A_1	$\$1.25 \times 10^6$	$\$1.25 \times 10^6$	$\$1.35 \times 10^6$
Annual Rate of Change of Labor Cost: $g\%$	3%	3%	3%

Solution: ***Find the Least-Cost Bridge***

Table 5.9 compares the capitalized costs of the cash flow elements of the *X*, *Y*, and *Z* bridges. As in Example 4.7, *the Z bridge is the least-cost alternative.*

TABLE 5.9 Capitalized Cost Comparison of Bridges *X*, *Y*, and *Z* When $ST = ST_\infty$ and $r^*\% = 7\%$ per Year

WORTH TYPE	BRIDGE CAPITALIZED COST TABLE		
	X	*Y*	*Z*
CC_P	$\$60 \times 10^6$	$\$65 \times 10^6$	$\$67 \times 10^6$

TABLE 5.9 Capitalized Cost Comparison of Bridges *X*, *Y*, and *Z* When $ST = ST_\infty$ and $r^*\% = 7\%$ per Year *(continued)*

Worth Type	Bridge Capitalized Cost Table		
	X	*Y*	*Z*
CC_A	$\frac{A}{r^*} = \frac{\$1.75 \times 10^6}{0.07}$ $= \$25 \times 10^6$	$\frac{A}{r^*} = \frac{\$1.50 \times 10^6}{0.07}$ $= \$21.4 \times 10^6$	$\frac{A}{r^*} = \frac{\$1.00 \times 10^6}{0.07}$ $= \$14.3 \times 10^6$
CC_G	$\frac{G}{r^{*2}} = \frac{\$0.100 \times 10^6}{0.07^2}$ $= \$20.4 \times 10^6$	$\frac{G}{r^{*2}} = \frac{0.075 \times 10^6}{0.07^2}$ $= 15.3 \times 10^6$	$\frac{G}{r^{*2}} = \frac{\$0.050 \times 10^6}{0.07^2}$ $= \$10.2 \times 10^6$
CC_{A_1}	$\frac{A_1}{r^*-g} = \frac{\$1.25 \times 10^6}{0.07-0.04}$ $= \$31.3 \times 10^6$	$\frac{A_1}{r^*-g} = \frac{\$1.25 \times 10^6}{0.07-0.04}$ $= \$31.3 \times 10^6$	$\frac{A_1}{r^*-g} = \frac{\$1.35 \times 10^6}{0.07-0.04}$ $= \$33.8 \times 10^6$
ΣCC	$CC_X = \$136.7 \times 10^6$	$CC_Y = \$133.0 \times 10^8$	$CC_Z = \$125.3 \times 10^8$

Consistency Check

The annual worth limit of each alternative is just its capitalized cost times r^*:

$$AW_{X,\infty} = CC_X \times r^* = \$136.7 \times 10^6 \times 0.07 = \underline{\underline{\$9.57 \times 10^6}}$$

$$AW_{Y,\infty} = CC_Y \times r^* = \$133.0 \times 10^6 \times 0.07 = \underline{\underline{\$9.31 \times 10^6}}$$

$$AW_{Z,\infty} = CC_Z \times r^* = \$125.3 \times 10^6 \times 0.07 = \underline{\underline{\$8.77 \times 10^6}}.$$

Except for the negligible one digit round-off error in the third digit of $AW_{Z,\infty}$, these annual worth limits are identical to those obtained in Example 4.7.

5.4 PRESENT WORTH INVESTMENT CALCULATIONS

This section is the *present worth* version of Section 4.6, Annual Worth Investment Calculations. The General Repayment Equation shows that the present worths of the receipts and disbursements of a *simple* investment are equal. Hence, we can make a variety of present worth *equivalence* calculations at a *specified* IRR. In all of them we solve for *one* unknown. The following list describes three common types of *equivalence* calculations at a *specified* IRR:

1. Calculate the metric of a receipt stream that yields a specified IRR.
2. Calculate the metric of a disbursement series that yields a specified IRR.
3. Find the counteroffer for an investment that yields a specified IRR.

EXAMPLE 5.5 Calculating the Required Annual Receipt of an Investment with a Present Worth Calculation (*E*-Class)

Problem: The equipment to produce a small automotive part costs \$125,000 (This is the *specified* amount *P* of the investment.) and has a salvage value *S* of \$6,000 at the end of 10 years. The *annual* disbursements A_D are \$3,000 per year. *(1) What annual receipt A_R from the sales of the automotive part is required to yield an IRR% of 12% per year? (2) Compare the result with that of Example 4.8.*

Solution: ***Calculate the Annual Receipt***

As with any simple investment, the present worths of the disbursements and receipts must be equal:

$$\underbrace{A_R(P/A,12\%,10)+S(P/F,12\%,10)}_{PW\text{ of Receipts}}=\underbrace{P+A_D(P/A,12\%,10)}_{PW\text{ of Disbursements}}.$$

Hence,

$$A_R=\frac{\underbrace{P}_{\$125{,}000}+\underbrace{A_D}_{\$3{,}000}\underbrace{(P/A,12\%,10)}_{5.6502}-\underbrace{S}_{\$6{,}000}\underbrace{(P/F,12\%,10)}_{0.32197}}{\underbrace{(P/A,12\%,10)}_{5.6502}}=\underline{\underline{\$24{,}781.}}$$

As expected, the result is identical to that of Example 4.8. Both calculations require two factor values, but this one has one more numerical operation (division).

EXAMPLE 5.6 Calculating a Counteroffer for a Pizza Parlor Investment with a Present Worth Calculation (Class 2E)

Problem: The owner of a pizza parlor wants to sell it for \$500,000 so that he can retire to Florida. His accountant certifies that the annual receipts A_R for the past two years were \$350,000 per year. The annual operating cost A_D for ingredients, beverages, utilities, maintenance, and labor is expected to be \$250,000 this year, but it has been increasing at a rate of 3% per year. Your uncle is interested in purchasing the pizza parlor, but he is concerned that the annual receipts are stagnant while the annual operating costs keep rising. Hence, he would like to earn 20% per year on the investment over the next nine years to account for this risk. *(1) What should his counteroffer P be if the "salvage" value of the pizza parlor in nine years is <u>assumed</u> to be P? (2) Compare the result with that of Example 4.9.*

Solution: ***Calculate the Value of P***

As with any simple investment, the present worths of the disbursement and receipt streams must be equal:

$$\underbrace{A_R(P/A,20\%,9)+S(P/F,20\%,9)}_{PW\text{ of Receipts}}=\underbrace{P+A_D(P/A_1,3\%,20\%,9)}_{PW\text{ of Disbursements}}.$$

Hence,

$$P-P\underbrace{(P/F,20\%,9)}_{0.19381}=\underbrace{A_R}_{\$350{,}000}\underbrace{(P/A,20\%,9)}_{4.0310}-\underbrace{A_D}_{\$250{,}000}\underbrace{(P/A_1,3\%,20\%,9)}_{4.3949},$$

and the value of P rounded to three significant digits is given by

$$P=\frac{\$350{,}000\times 4.0310-\$250{,}000\times 4.3949}{1-0.19381}=\underline{\underline{\$387{,}000.}}$$

This result is identical to that of Example 4.9. Both calculations require three factor values, but this one has one more numerical operation (multiplication).

5.5 CHOOSING WORTH CALCULATION METHODS

5.5.1 Least-Cost Studies

Table 5.10 lists the advantages of the annual and present worth cost comparison methods. It leads to the following:

Usage Note: *An annual worth comparison is the best way to compare cost alternatives and present the results to an audience without a background in engineering economics.*

TABLE 5.10 Advantages of the Annual and Present Worth Cost Comparison Methods

ADVANTAGE	UNIFORM WORTH METHOD	PRESENT WORTH METHOD
Understandable Without a Background in Engineering Economics	YES	NO
Simpler Worth Comparison for Irregular Alternatives	YES[1]	YES

[1] A present worth comparison can easily be converted to an annual worth comparison by means of Equation (4-1).

5.5.2 Other Calculations

Present worths are customarily used to

1. *Calculate the IRR of a simple investment* because the equality of the present worths of the receipt and disbursement series leads to the General Polynomial Equation (GPE) for finding the unique IRR of a simple investment. Excel's IRR function quickly makes this calculation.
2. *Calculate the unknown in equality calculations* when the cash flows are *irregular*. Using uniform worths for these calculations requires the *additional* step of annualizing the present worths. *When the streams are regular either worth method can be used.*

5.6 SUMMARY

Example 5.1 in Subsection 5.1.1 illustrates the use of a present worth table to make a least-cost study on a *specified* service time of *two or more* cost alternatives without replacement costs. The ***PRESENT WORTH COMPARISON Calculator*** is used in Subsections 5.1.2 and 5.1.3 to automate the calculations. Subsection 5.1.4 shows how to incorporate the *replacement* costs of a component of each cost alternative into a least-cost study, and Example 5.2 illustrates the procedure. The ***PRESENT WORTH COMPARISON Calculator with Component Replacement*** in Subsection 5.1.5 automates the calculations.

Subsection 5.2.1 defines **extra capacity** alternatives, and Example 5.3 shows how to incorporate one into a present worth least-cost study. Subsection 5.2.2 explains that **dissimilar** alternatives are calculated just like we calculate similar alternatives. However, calculating the present worths of **irregular** alternatives is simpler than calculating their annual worths because the first step in doing so is the calculation of their present worths.

Section 5.3 explains how to compare two or more cost alternatives in ***perpetual*** service and illustrates the procedure with Example 5.4. Section 5.4 shows that we can make various types of present worth *equivalence* calculations for investments with a *specified* IRR. In all of them we solve for *one* unknown. Examples 5.5 and 5.6 illustrate these types of calculations.

Section 5.5 explains how to choose between annual worth and present worth calculations.

5.7 PROBLEMS

(P5-1) *Redo problem (P4-1) using a present worth comparison of the cost alternatives.*

(P5-2) *Redo problem **(P4-2)** using a present worth comparison of the cost alternatives.*

(P5-3) *Redo problem **(P4-3)** using a present worth comparison of the cost alternatives.*

(P5-4) *Redo problem **(P4-4)** using a present worth comparison of the cost alternatives.*

(P5-5) *Redo problem **(P4-5)** using a present worth analysis of the investment.*

(P5-6) *Which worth comparison method do you prefer? Why?*

6

Simple and Quasi-Simple Investments

OBJECTIVES

After reading this chapter, you will be able to

1. Calculate the *pretax* IRR%s of a simple manufacturing plant investment and a simple cost reduction investment, ***SI*-Class** problems.
2. Determine if a multi-simple investment is a quasi-simple investment.
3. Calculate the *pretax* IRR%s of two types of quasi-simple investments: bi-simple and tri-simple investments, ***QSI*-Class** problems.

The effect of income taxes is explained in Chapter 9.

6.1 SIMPLE INVESTMENTS: A QUICK REVIEW

6.1.1 Definition

Simple investments were explained in Subsection 2.9.3. A **simple investment** has one or more *initial* negative cash flows followed by one or more positive cash flows:

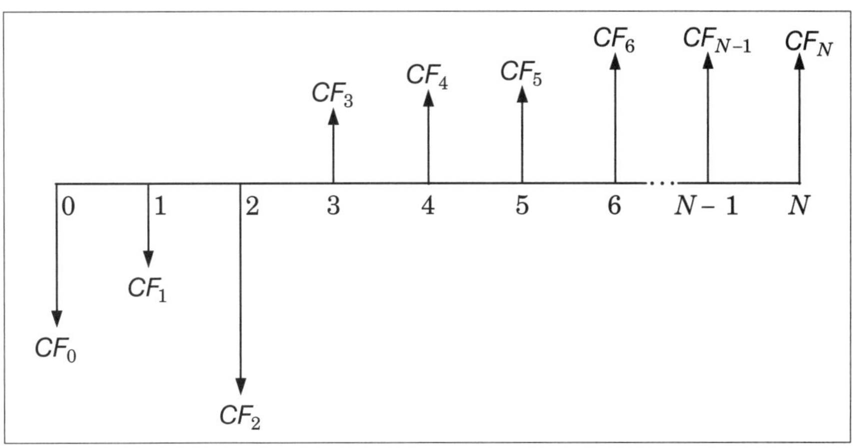

Figure 6.1 **CFD of a Simple Investment**

6.1.2 Properties

Unique Rate of Return (ROR)

A simple investment has a unique ROR because it has only one change in the signs of its cash flows. It is obtained from the General Polynomial Equation (GPE):

$$CF_0x^N + CF_1x^{N-1} + \cdots + CF_{N-1}x + CF_N = 0, \qquad (2\text{-}111)$$

where $i^* \equiv x_+ - 1$ and x_+ is the *unique* positive root of Equation (2-111). This unique ROR is called the **internal rate of return** (IRR) to emphasize that it is generated by the investment and is not specified like the interest rate of a loan. We normally calculate the IRR% of a simple investment with Excel's IRR function. Section 2.10 explains how this is done. We used the IRR function in Example 2.22 to calculate the IRR% of a simple investment and in Example 3.4 to calculate a *monthly* IRR% (of a spec house) that we converted to an *APY*. Lastly, we used the IRR function in Example 3.5 to determine a counteroffer for a machine shop investment.

Positive Intermediate Balances

Here is the proof that the intermediate balances of a simple investment are positive. If the last *negative* cash flow of a simple investment occurs at the *end* of the Kth interest period, we use the future worth form of the Intermediate Balance Equation (IBE), Equation (3-2), to calculate the intermediate balances at the ends of interest periods $1, 2, \ldots, K$:

$$B_{k^*} = \sum_{k=0}^{k^*} \left(P_k - R_k\right)(1+i)^{k^*-k} \qquad k^* = 1, 2, \ldots, K. \qquad (3\text{-}2)$$

Since all the cash flows, $CF_k \equiv R_k - P_k$, up to the end of the Kth interest period are negative,

$$B_k > 0 \qquad k = 1, 2, \ldots, K. \qquad (6\text{-}1)$$

We use the present worth form of the IBE, Equation (3-4),to calculate the intermediate balances at the ends of interest periods $K+1, K+2, \ldots, N-1$:

$$B_{k^*} = \sum_{k=k^*+1}^{N-1} \frac{R_k - P_k}{(1+i)^{k-k^*}} \qquad k^* = K{+}1, K{+}2, \ldots, N-1. \qquad (3\text{-}4)$$

Since all the cash flows *after* the end of the Kth interest period are positive,

$$B_k > 0 \qquad k = K+1, K+2, \ldots, N-1. \qquad (6\text{-}2)$$

It follows from Equations (6-1) and (6-2) that

$$\boxed{B_k > 0 \qquad k = 1, 2, \ldots, N-1.} \qquad (6\text{-}3)$$

Hence, ***the intermediate balances of a simple investment are positive.***

6.2 IRR% of a MANUFACTURING PLANT INVESTMENT & END-of-YEAR CONVENTION

This section explains the disbursements required for an investment in a manufacturing plant that will be constructed to produce consumer and/or industrial products. It explains the rationale for the End-of-Year Convention and illustrates it by calculating the investment's IRR%.

6.2.1 Disbursement Classes

An investment in the construction of a manufacturing plant has **three classes of disbursements**: capital, operating, and income tax.

Capital Disbursements

There are **three types of capital disbursements**:

- **Land Capital** LC is the total disbursements for the plant site. It consists of the costs of the land, surveys, legal services, and testing to determine the load-bearing and drainage capacities of the site soil.
- **Fixed Capital** FC is the total disbursements for the design, construction, and *startup* of the plant and the purchase of its equipment, control system, piping, ducting, electrical cables and wiring, structural steel for pipe racks, ladders, and walkways, and its paint and insulation.
- **Working Capital** WC is the *additional* money invested to create adequate product and raw material inventories and assure the *timely* payment of salaries and invoices—such as those for utilities, raw materials, and supplies—during the operation of the plant.[1] *At the end of the investment's life, the working capital is recovered!*

The **total capital** TC is the sum of the three capitals:

$$\boxed{TC \equiv LC + FC + WC.} \tag{6-4}$$

Operating Disbursements

There are **two types of operating disbursements** OD—production costs and SCORD costs. There are **four categories of production costs** PC:

- **Material—**raw materials and supplies
- **Utility—**steam, refrigeration, electricity, water, and fuel
- **Investment—**repairs and maintenance, property taxes, insurance premiums, and security
- **Labor—**salaries and benefits, such as vacation and sick days, medical insurance, and pension contributions, for the operators, supervisors, and support personnel

There are **three categories of SCORD costs** SC:

- **Sales**—salaries and benefits for the sales personnel and the costs to maintain their offices

[1] It's analogous to the money that must always be available in the cash register and bank account of a pizzeria to maintain an adequate supply of dough, toppings, and beverages and the timely payment of utility bills, insurance premiums, and salaries.

- **Corporate Overhead**—salaries and benefits for the executive, financial, engineering, and legal personnel and the costs to maintain their offices
- **Research & Development**

Practice Note: *Small businesses have limited SCORD costs. They usually don't have corporate overhead and research and development costs.*

6.2.2 Cash Flow Timing

The capital disbursements occur in the **capital time** CT, and the operating and income tax disbursements occur in the **operating time** OT. The investment's **life** L is the sum of these times:

$$\boxed{L = CT + OT.} \tag{6-5}$$

Capital Time

The three capitals are disbursed as follows:

1. The land capital is disbursed at the *beginning* of the capital time.
2. The fixed capital is disbursed *throughout* the capital time.
3. The working capital is disbursed at the *end* of the capital time.

Operating Time

Operating disbursements and receipts occur *throughout* each year of the operating time, and income taxes are paid *quarterly* in each of those years.

At the end of the investment's life, three events occur:

1. The working capital is recovered.
2. The plant is sold for its salvage or market value.
3. The land is sold.

6.2.3 End-of-Year Convention

Except for P_0 and R_0, the cash flows of a loan are *assumed* to occur at the ends of interest periods. For example, when the loan is a bank account, the interest period is normally one day. This insures that a deposit or a withdrawl is always recognized at the end of the day in which it occurs. The interest period of most consumer loans, such as automobile loans and home mortgages, is one month. There is a 15 to 20 day time frame in which the borrower can make a payment that will be credited to the end of the interest period. Failure to do so, normally results in a late charge.

Definition

When a plant is built, capital disbursements occur irregularly, but receipts and operating disbursements occur with some regularity throughout a year. However, unlike the cash flows of a bank account or an auto loan, *the estimated cash flows of an investment are not normally recognized monthly, daily, or continuously because the uncertainties in their estimated values do not justify such precise models.* Hence, we use a cash flow model that makes it easy to calculate an IRR% with an accuracy that is commensurate with the *estimated values* of the key economic variables.

The model normally used to recognize the transfers of receipts and disbursements is called the **End-of-Year Convention** (EoY Convention):

End-of-Year Convention
(1) The interest period is one year because the accounting, budgeting, and income tax time frames of businesses are each one year.
(2) Except for the purchase of a plant site or an existing plant, the total capital disbursements in a year occurs at the *end* of it.
(3) The total receipts, operating disbursements, and income taxes in a year occur at the *end* of it.

Computation Note: *The EoY Convention is an economic model that simplifies the calculation of the IRR%, but it normally does not compromise an investment decision based on it.*

EoY Convention for a Capital Time with a Fixed Capital S-Curve

The fixed capital cash flows for the construction of a manufacturing plant have an *atypical* pattern. A large amount of relatively low-cost engineering time is required before the major disbursements for equipment, materials, and construction can occur. Therefore, the variation of the ***cumulative*** **fixed capital cash flow** over the capital time has the shape of an **S-curve**, such as the one in Figure 6.2.

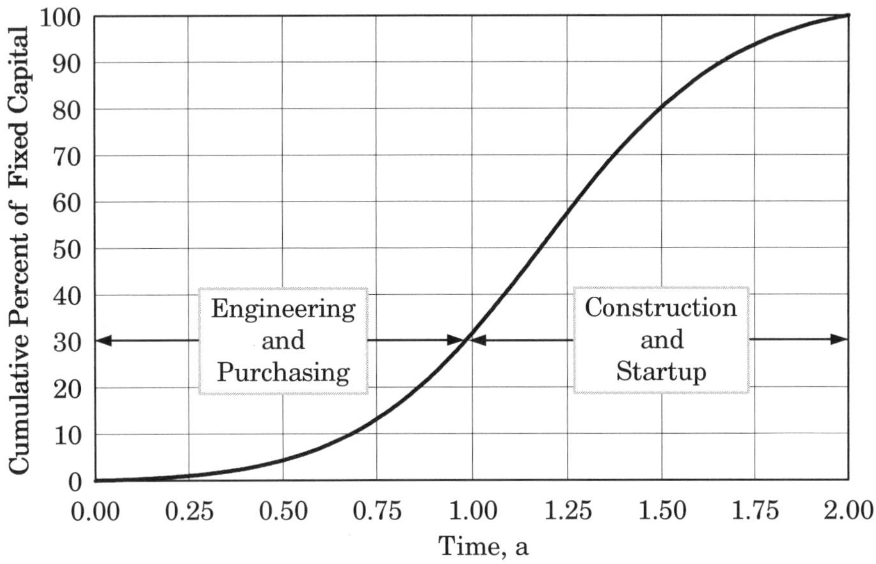

Figure 6.2 **S-Curve of Cumulative Fixed Capital Cash Flow** NOTE: A longer capital time or a different type of project would have a *different* S-Curve.

Engineering and Purchasing

The inexpensive but time-consuming engineering *and* the resulting purchasing activities consume approximately 50% of the capital time and 30% of the fixed capital. Vendor bids for the supply of equipment can't be solicited and analyzed *until* equipment specifications and drawings have been prepared by the project's engineers and drafting personnel, respectively. After successful vendors are selected, the equipment for the plant is purchased. *This accounts for approximately 20% of the fixed capital.* In addition to these engineering and purchasing activities, various types of engineering drawings—foundation, structural steel, electrical, piping, and plant layout—must be prepared for the construction phase of the project.

Construction and Startup

Construction begins with the clearing and grading of the plant site; then equipment foundations are built. Equipment and materials, such as structural steel, piping, gas ducts, and electrical cables, are sequentially delivered to the site and the plant is constructed. After a successful startup, it begins to generate sales of its product.

Computation Note: *(1) Based on the preceding description of the S-curve, we will often assume that one third of the fixed capital is disbursed at the end of the first year and the remainder is disbursed at the end of the second year.*
(2) This distribution of fixed capital must be adjusted for longer capital times.

EoY Convention for the Operating Time

Receipts and operating disbursements tend be *uniformly* distributed throughout a year, and income taxes are paid quarterly in it. Hence, *it is conservative to assume that the total receipts, operating disbursements, and income taxes occur at the end of each year in the operating time. This minimizes the present worths of the positive cash flows and, hence, the estimated IRR%.*

Practice Note: *The estimates of future receipts and disbursements are the most uncertain quantities in any prediction of the IRR% of a proposed investment in a manufacturing plant because one or more unfavorable economic events may occur:*

- A competitor may offer the *same* product at a *lower* price. *This will result in lower receipts.*
- *Energy and/or raw material costs may rapidly increase* because a Middle East crisis reduces the supply of imported oil.
- A competitor may offer a product with *better* features at the *same* price. *This will significantly reduce receipts or may even result in a plant closure. The latter event will reduce the assumed operating time!*
- The plant may be forced to shut down because it is not in compliance with environmental regulations or its product is identified as a health hazard. *This too will reduce the assumed operating time!*

Limitations of the EoY Convention

There are times when the EoY Convention can lead to an inaccurate IRR%. This occurs when (1) an investment's life is a few years or (2) there are a few irregular disbursements and/or receipts in each year of the investment's life. A common example of the first case is an investment to build a spec house. We use an interest period of one month and convert the calculated monthly IRR% to an annual percentage yield. Example 3.4 illustrates this type of calculation.

6.2.4 *Pretax* IRR% of a Chemical Plant Investment (*SI*-Class)

To illustrate the EoY Convention, let's calculate the *pretax* IRR% of a *proposed* chemical plant investment.[2] The *estimated* values of the economic variables are listed in Table 6.1, and Figure 6.3 is the CFD of this *simple* investment.

[2] *The effect of income taxes on the IRR% is explained in Chapter 9.*

TABLE 6.1 *Estimated* Values of the Economic Variables of a *Proposed* Chemical Plant Investment[1]

ECONOMIC VARIABLE	*ESTIMATED* VALUE
Fixed Capital: FC	$\$69.6 \times 10^6$
Land Capital: $LC \approx 0.01FC$	$\$0.696 \times 10^6$
Working Capital: $WC \approx 0.15FC$	$\$10.44 \times 10^6$
Salvage Value: $S \approx 0.05FC$	$\$3.48 \times 10^6$
Annual Receipt: R	$\$208.8 \times 10^6$
Annual Production Cost: PC	$\$147 \times 10^6$
Annual SCORD Cost: $SC \approx 0.15R$	$\$31.32 \times 10^6$
Capital Time: CT	2 a
Operating Time: OT[2]	10 a

[1] The symbol ≈ in the table indicates that the expression is a *cost estimating* equation.
[2] The operating time is a *conservative* value. It is often larger in practice.

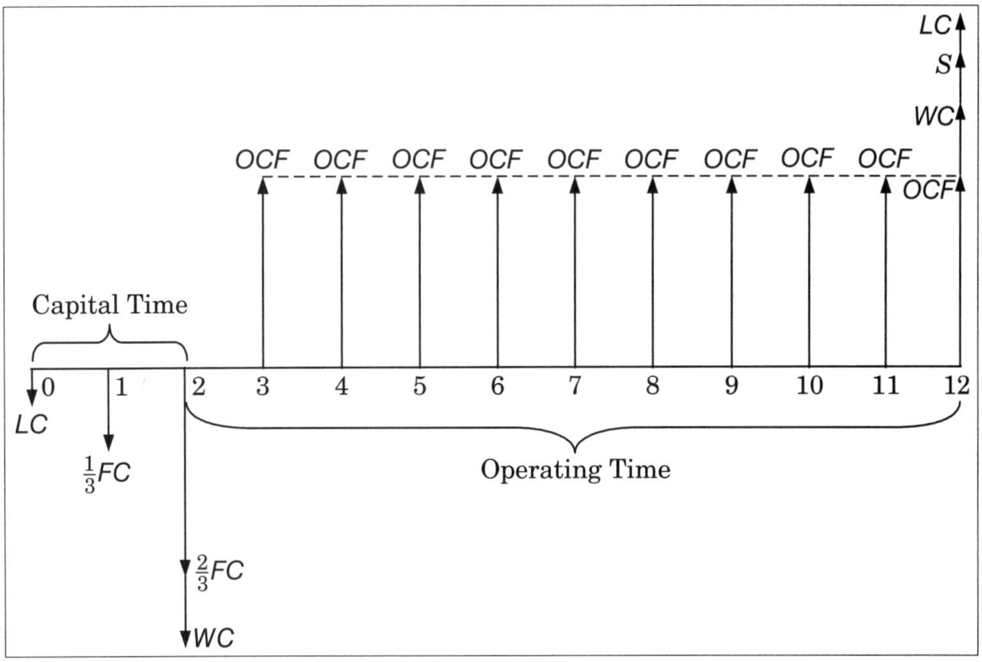

Figure 6.3 ***Pretax* CFD of the Chemical Plant Investment**

The **operating cash flow** OCF_k in the kth year of *any* manufacturing investment is the difference of the annual receipt and the annual production and SCORD costs:

$$OCF_k \equiv R_k - (PC_k + SC_k). \tag{6-6}$$

For the chemical plant investment,

$$\begin{aligned} OCF_k &= R_k - (PC_k + SC_k) \\ &= 208.8 \times 10^6 - (147 + 31.32) \times 10^6 = 30.48 \times 10^6 \qquad k = 3, 4, \ldots 12. \end{aligned}$$

Note the following in the CFD:

1. One third and two thirds of the fixed capital *FC* are disbursed at the ends of the first and second years, respectively.
2. The working capital *WC* and land capital *LC* are recovered at the end of the operating time. The salvage value *S* is also received at this time.
3. It is *assumed* that *the land is sold for its original price* because we don't want a *potentially* spectacular increase in land values to distort the value of the IRR% for the underlying chemical plant investment.

Spreadsheet 6.1 calculates a *pretax* IRR% of 32.59% per year.[3]

	A	B	C	D	E
1					
2		***Pretax IRR% of a CHEMICAL PLANT INVESTMENT: CT = 2a***			
3			**FC**	$69,600,000	=69.6*10^6
4			**LC**	$696,000	=0.01*D3 These are typical
5			**WC**	$10,440,000	=0.15*D3 ***cost-estimating***
6			***S***	$3,480,000	=0.05*D3 ***equations.***
7			***R***	$208,800,000	=208.8*10^6
8			**PC**	$147,000,000	= 147* 10^6
9			**SC**	$31,320,000	=0.15*D7 This is a typical ***cost-estimating equation.***
10			***k***		**Cash Flow CF_k, $$10^6$**
11			0	$-LC$	(0.696)
12			1	$-\frac{1}{3}FC$	(23.2)
13			2	$-\left(\frac{2}{3}FC + WC\right)$	(56.84)
14			3		30.48
15			4		30.48
16			5		30.48
17			6		30.48
18			7	$OCF = R - (PC + SC)$	30.48
19			8		30.48
20			9		30.48
21			10		30.48
22			11		30.48
23			12	$OCF + LC + WC + S$	45.096
24				**=IRR(E11:E23)**	**32.59%**

Spreadsheet 6.1 ***Pretax IRR% of a CHEMICAL PLANT INVESTMENT:CT = 2a***
NOTES: (1) The spreadsheet is located in the *Book Templates* file. (2) Its inputs are the values of *FC*, *R*, and *PC*.

[3] Spreadsheet 6.1 can be used for other types of manufacturing plant investments by modifying one or more of the cost estimating equations and *possibly* the value of *CT*.

6.3 SIMPLE COST REDUCTION INVESTMENTS

6.3.1 Definition

Individuals, families, and companies are always looking for ways to reduce the cost of an activity:

- A salesman can reduce the fuel and insurance costs of his automobile by *replacing* it with a less expensive one with better fuel economy.
- A couple can reduce the heating and cooling bills of their home by *replacing* its old single-pane windows with modern dual-pane units.
- A company can reduce the cost of detecting defective products by *replacing* the two workers that perform this task with a robot.

In each case *a fixed capital disbursement is required to achieve the cost reduction.* Hence, we define a **cost reduction investment** as one or more capital disbursements that reduce the operating costs of an activity.

6.3.2 Calculating IRR% (*SI*-Class)

Computation Notes: *(1) The cash flows of a cost reduction investment are the* ***differences*** *of the cost reduction cash flows and those of the existing activity.*
(2) If the cost reduction investment is a simple investment, we calculate its IRR% like we calculate that of any simple investment.

For example, the differences of the yearly *Make* and *Buy* cash flows of the alternatives in Example 4.6 constitute a simple cost reduction investment. Spreadsheet 6.2 calculates its *IRR%* as 26.91% per year.

	A	B	C	D	E
1					
2		***IRR% of an iPod ENGRAVING COST REDUCTION INVESTMENT***			
3		***k***	***Buy*** **Cash Flows, $**	***Make*** **Cash Flows, $**	***Make*** **minus** ***Buy*** **Cash Flows, $**
4		0	0	(27,000)	(27,000)
5		1	(21,000)	(11,400)	9,600
6		2	(23,625)	(12,825)	10,800
7		3	(23,100)	(12,540)	10,560
8		4	(22,225)	(12,065)	10,160
9		5	(26,250)	(14,250)	12,000
10				**=IRR(E4:E9)**	**26.91%**

Spreadsheet 6.2 ***IRR% of an iPod ENGRAVING COST REDUCTION INVESTMENT*** NOTES: (1) The *fixed* capital investment is the purchase of the $27,000 engraving machine. (2) The cash flows of the *Make* –*Buy* cost reduction investment are the differences of the *Make* cash flows and the *Buy* cash flows. (3) This cost reduction investment is a simple investment.

6.4 MULTI-SIMPLE INVESTMENTS

6.4.1 Definitions, Properties & Occurrences

The cash flow pattern of a **multi-simple investment** is *a sequence of two or more simple investment patterns.* A **bi-simple investment** is a sequence of *two* simple investment patterns, and a **tri-simple investment** is a sequence of *three* simple investment patterns. Figure 6.4 is a CFD of a tri-simple investment.

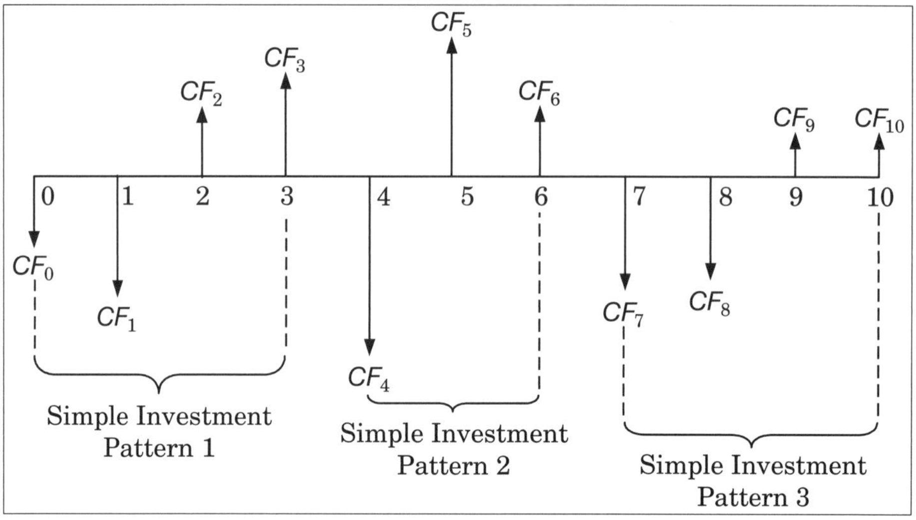

Figure 6.4 **CFD of a Tri-Simple Investment** Note: There are two negative cash flows in the first and third simple investment patterns.

Bi-simple and tri-simple investments are the common multi-simple investments. A bi-simple investment occurs when a simple investment has an *unexpected* negative cash flow where a positive one had been predicted. For example, a recession or a new competitor may cause such a large decline in a company's annual sales that its annual cash flow becomes negative. Another example of bi-simple investment is a spec house that is sold in the *middle* of the construction phase. The down payment, which is usually 10% of the sales price, yields a *positive* cash flow *preceded* by a series of negative cash flows and *followed* by another series of negative cash flows until the house is finished and sold; this yields another positive cash flow.

A tri-simple investment occasionally occurs when we want to calculate the IRR% of the ***incremental investment*** *of* ***two mutually exclusive investments (MEIs)*** *(see Chapter 8).*

The cash flows of a multi-simple investment with N simple investment patterns have an **odd number of sign changes** δ_N:

$$\delta_N = 2N - 1. \tag{6-7}$$

Hence, it follows from Descartes' Rule of Signs that a bi-simple investment may have *one* or three *positive* roots of the GPE, and a tri-simple investment may have *one,* three, or five.

6.4.2 IRR of a Quasi-Simple Investment (*QSI*-Class)

As the negative cash flows *after* the first pattern of a multi-simple investment approach zero, it becomes a simple investment. But even for *finite* values of these negative cash flows, a multi-simple investment *may* behave like a simple investment because it can have just *one* positive root of the GPE:

Computation Notes: *(1) If the GPE has **one** positive root x_+, the value of i^* is **unique**.*
*(2) If the intermediate balances calculated with i^* are **positive**, i^* equals the IRR of the multi-simple investment, which we call a **quasi-simple investment**.*
*(3) If there is at least one negative intermediate balance calculated with i^*, the multi-simple investment is a **non-simple investment**.*

The various types of non-simple investments, including bi-simple and tri-simple investments, are explained in Chapter 7.

We use Step-by-Step 6.1 to determine if a multi-simple investment has a *unique* value of i^* and *positive* intermediate balances:

Multi-Simple-Quasi-Simple Investment Test:
Step-by-Step 6.1

(1) To see if there is only *one* positive root of the GPE, find its roots with a graphing calculator, mathematics software, or an Internet polynomial root finder.

(2A) If there is only *one* positive root x_+, calculate i^* with Equation (2-110):

$$i^* = x_+ - 1 \qquad (2\text{-}110)$$

and proceed to Step 3.

(2B) If there are ***three*** or more positive roots, **Stop!** The multi-simple investment is a **non-simple investment.** *Go to Section 7.5.*

(3) Use the Recursive Balance Function, Equation (3-11), to *recursively* calculate the *intermediate* balances in Excel:

$$B_k = B_{k-1}(1 + i^*) + CF_k \qquad k = 1,2,\ldots,N. \qquad (3\text{-}11)$$

(4A) If *all* the intermediate balances are *positive*, i^* is the IRR. Hence, the multi-simple investment is a **quasi-simple investment**.

(4B) If there is at least *one negative* intermediate balance, the multi-simple investment is a **non-simple investment.** *Go to Section 7.5.*

Computation Notes: *(1) We can't use Excel to determine if there is a **unique** value of i^* because its IRR function only searches for **one** root of the GPE. If there are multiple roots it does not try to find them!*
(2) Mathcad and MATLAB have very convenient polynomial root finders.

EXAMPLE 6.1 Determining if a Bi-Simple Investment Is a Quasi-Simple Investment (*QSI*-Class)

Problem: A deep recession resulted in an unexpected *negative* cash flow in year 5 of a tractor factory investment:

CASH FLOW, $10^6	**−15**	**−15**	15	20	25	**−10**	15	10	5	5	5	5
YEAR *k*	0	1	2	3	4	5	6	7	8	9	10	11

The negative value of CF_5 makes this a *bi-simple* investment. *Is it a quasi-simple investment?* That is, does it have a *unique i** and *positive* intermediate balances?

Solution: We first determine if the GPE has a unique positive root. To do this, we can use Mathcad[4], MATLAB, or any other software that has a *convenient* polynomial root finder. The Mathcad calculation below shows that there is a unique positive root of the GPE: 1.35772.

$$CF := \begin{pmatrix} 5 \\ 5 \\ 5 \\ 5 \\ 10 \\ 15 \\ -10 \\ 25 \\ 20 \\ 15 \\ -15 \\ -15 \end{pmatrix} \qquad polyroots\left(CF\right) = \begin{pmatrix} -1.63981 \\ -0.67917 \\ -0.55300 + 0.95548i \\ -0.55300 - 0.95548i \\ -0.39287 - 0.56747i \\ -0.39287 + 0.56747i \\ 0.26507 - 0.70177i \\ 0.26507 + 0.70177i \\ 0.66142 - 0.48709i \\ 0.66142 + 0.48709i \\ 1.35772 \end{pmatrix}$$

Computation Note:*Making the column vector CF of the coefficients of the GPE the argument of the polyroots ( ) function returns the column vector of the roots of the GPE. NOTE: The first coefficient* CF_0 *of the GPE is the* <u>*bottom*</u> *component of* **CF**.

It follows from Equation (2-110) that

$$i^{*}\% = 100(x_{+} - 1) = 100(1.35772 - 1) = 35.77\% \text{ per year.}$$

Use Excel to determine if the intermediate balances are positive. Spreadsheet 6.3 does this.

EXAMPLE 6.2 Determining if a Tri-Simple Investment Is a Quasi-Simple Investment (*QSI*-Class)

Problem: The cash flows of an investment are given below. The negative values of CF_4 and CF_8 make this a *tri-simple* investment. *Is this tri-simple investment a quasi-simple investment?*

CASH FLOW, $10^6	−8	4	5	6	**−3**	6	7	8	**−4**	4	3	2
YEAR *k*	0	1	2	3	4	5	6	7	8	9	10	11

[4] PTC Mathcad is a registered trademark of Parametric Technology Corporation.

	A	B	C	D	E
1					
2		***IRR% and BALANCES of a QUASI-SIMPLE BI-SIMPLE IINVESTMENT***			
3		***k***	CF_k, **\$10**^{6}	$B_k = B_{k-1}(1+i^*) - CF_k$	**Unique** $i^*\%$
4		0	(15)	15	35.77%
5		1	(15)	35.3658	= D4*(1 + C$16) − C5
6		2	15	33.0167	Copy Cell D5 and Paste it Down Array D6:D15
7		3	20	24.8274	
8		4	25	8.7086	
9		5	(10)	21.8238	
10		6	15	14.6306	
11		7	10	9.8642	
12		8	5	8.3928	
13		9	5	6.3950	
14		10	5	3.6827	
15		11	5	0.0000	
16			**35.77%**	**=IRR(C4:C15)**	

Spreadsheet 6.3 *Excel's recursive calculation of the intermediate balances shows that they are all positive.* Hence, ***this bi-simple investment is a quasi-simple investment.*** NOTES: (1) The value of $i^*\%$ is calculated as a check of the Mathcad result in cell E4. (2) Label cell D3 shows that each cell in formula array D5:D15 is calculated with the Recursive Balance Function, Equation (3-11) in Step-by-Step 6.1. (3) The balance formulas in array D6:D15 are easily obtained by copying the formula in cell D5 and pasting it down the array.

Computation Note:*The final balance B_N is* ***always*** *zero because this was a specified condition in the derivation of the General Repayment Equation (GRE), Equation (2-37), and, hence, the General Polynomial Equation (GPE), Equation (2-111), derived from it.*

Solution: A polynomial root finder shows that the unique positive root of the GPE is 1.49792.[5] Hence,

$$i^*\% = 100(x_+ - 1) = 100(1.49792 - 1) = 49.79\% \text{ per year.}$$

Use Excel To Determine If the Intermediate Balances Are Positive

Spreadsheet 6.4 shows that the intermediate balances are positive, and, as required, the final balance B_{11} is zero. Hence, ***this tri-simple investment is a quasi-simple investment.***

[5] *Even though a multi-simple investment may have multiple positive roots of the GPE, it is not unusual for it to have just one. However, even when this is so, one or more negative intermediate balances are not uncommon. This is illustrated in Section 7.5.*

	A	B	C	D	E
1					
2		***IRR% and BALANCES of a QUASI-SIMPLE TRI-SIMPLE INVESTMENT***			
3		***k***	***CF*** $_k$**, \$10**^{6}	$B_k = B_{k-1}(1+i^*) - CF_k$	**Unique** i^***%**
4		0	(8)	8	49.79%
5		1	4	7.9833	= D4*(1 + C$16) – C5
6		2	5	6.9583	Copy Cell D5 and Paste it Down Array D6:D15
7		3	6	4.4230	
8		4	(3)	9.6253	
9		5	6	8.4179	
10		6	7	5.6093	
11		7	8	0.4022	
12		8	(4)	4.6025	
13		9	4	2.8941	
14		10	3	1.3352	
15		11	2	0.0000	
16			**49.79%**	**= IRR(C4:C15)**	

Spreadsheet 6.4 *Excel's recursive calculation of the intermediate balances shows that they are all positive and that the final balance* B_{11} *is zero.* Hence, ***this tri-simple investment is a quasi-simple investment.***

6.5 CREATING a POLYNOMIAL ROOT FINDER in *MATHEMATICA*

6.5.1 Creating the *polyRoots* Function

You can't use Excel's IRR function to conveniently determine if there is a unique IRR% because it's a *one-trick pony.* That is, it only looks for the value of the IRR% that is closest to the default guess value of 10% or a user-specified guess value. But as mentioned previously, PTCMathcad and MATLAB have convenient polynomial root finders whose arguments are *lists* of the investment's cash flows. However, *if no convenient polynomial root finder is on your school or business computer and* ***Mathematica***[6] *is, you can use its* **NSolve** *function to find the roots of any polynomial.* But, unlike MathCad and MATLAB, ***its argument must be the polynomial itself!*** However, I created a polynomial root finder in *Mathematica* called ***polyRoots*** whose argument ***cashFlows*** *is* the list of the investments cash flows.

The ***polyRoots*** function is in the small *Mathematica* file, **PolyRoots FUNCTION.nb**, provided with this book. Here is what you see when you open *Mathematica* and click on the **PolyRoots FUNCTION.nb** file:

[6]Wolfram Research, Inc., Mathematica, Version 10.0, Champaign, IL (2014).

```
polyRoots[cashFlows_] :=
 Module[{},
  xPowers =
   Reverse[Table[x^i, {i, 0, Count[cashFlows, _] - 1}]];
  terms = cashFlows * xPowers;
  GPE = Sum[terms[[i]], {i, Count[cashFlows, _]}];
  NSolve[GPE, x]
 ]
```

The input variable **cashFlows** is a *list* of the investment's cash flows. To activate the **polyRoots** function, just *click anywhere in it and press ENTER on the numeric keypad or SHIFT plus ENTER:*

```
In[1]:= polyRoots[cashFlows_] :=
         Module[{},
          xPowers =
           Reverse[Table[x^i, {i, 0, Count[cashFlows, _] - 1}]];
          terms = cashFlows * xPowers;
          GPE = Sum[terms[[i]], {i, Count[cashFlows, _]}];
          NSolve[GPE, x]
         ]
```

6.5.2 Using the polyRoots Function

To illustrate the use of the **polyRoots** function, let's find the roots of the GPE in Example 6.2. Here is the list of the cash flows ***starting with the first cash flow*** CF_0. NOTE: ***The elements of a list must be enclosed by braces*** **{}**. *Click anywhere in the list and press ENTER on the numeric keypad or SHIFT plus ENTER:*

```
In[2]:= cashFlows = {-8, 4, 5, 6, -3, 6, 7, 8, -4, 4, 3, 2}

Out[2]= {-8, 4, 5, 6, -3, 6, 7, 8, -4, 4, 3, 2}
```

We now make **cashFlows** the argument of the **polyRoots** function and *press ENTER on the numeric keypad or SHIFT plus ENTER.* As expected, *the* ***unique*** *real root is identical to that in Example 6.2: 1.49792.*

```
In[3]:= polyRoots[cashFlows]

Out[3]= {{x → -0.870415 - 0.488478 i},
         {x → -0.870415 + 0.488478 i},
         {x → -0.458704 - 0.892036 i},
         {x → -0.458704 + 0.892036 i},
         {x → -0.320999 - 0.398516 i},
         {x → -0.320999 + 0.398516 i},
         {x → 0.492201 - 0.799233 i},
         {x → 0.492201 + 0.799233 i}, {x → 0.65896 - 0.5362 i},
         {x → 0.65896 + 0.5362 i}, {x → 1.49792}}
```

I could save the `polyRoots` function as a *Mathematica* function and call it the next time I want to use it. However, I just want to provide the minimum *Mathematica* knowledge required for any reader who has not used it before.

6.6 INCREMENTAL INVESTMENT CRITERION

An engineer tells her supervisor that a *least-cost* alternative Y has the *larger* purchase price. He will not accept this alternative unless the value of *IRR% for the incremental investment of Y and X exceeds the MARR%.* Let's see if this is true.

The cash flows of the **incremental investment** of Y and X are defined as the *differences* of their cash flows on their service time:

$$CF_{(Y-X),k} \equiv CF_{Y,k} - CF_{X,k} \qquad k = 0, 1, 2, \ldots, ST, \tag{6-8}$$

where *disbursement cash flows are defined as* ***negative*** *cash flows.*

When $Y - X$ is a *simple* incremental investment, a plot of PW_{Y-X} versus $r^*\%$ has the form displayed in Figure 6.5. It shows that when

$$r^*\% < i^*_{Y-X}\%, \tag{6-9}$$

alternative Y is always the least-cost alternative. Hence, *choosing Y also satisfies the requirement that the IRR% of the incremental investment $i^*_{Y-X}\%$ exceeds $r^*\%$.*

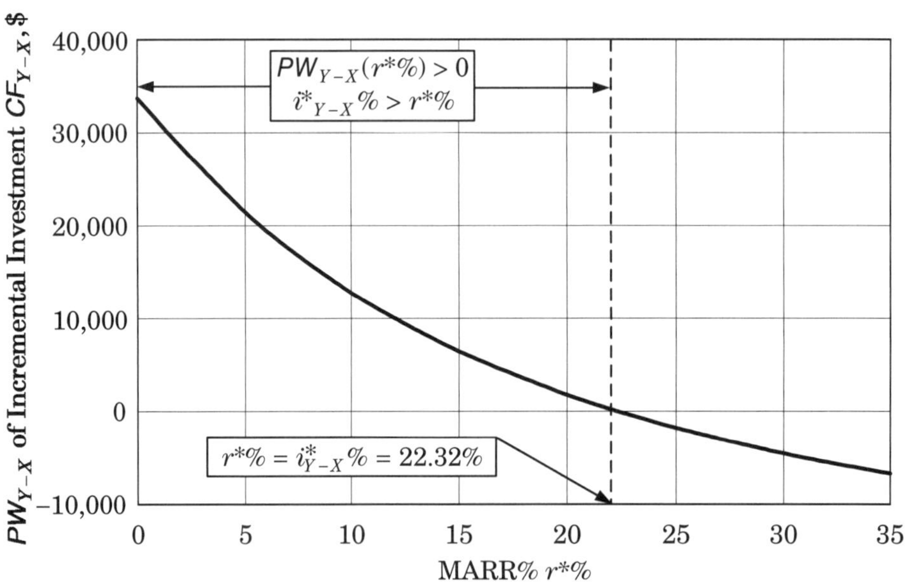

Figure 6.5 **Variation of the Present Worth of the *Simple* Incremental Investment $Y - X$ with $r^*\%$**

When

$$r^*\% = i^*_{Y-X}\%, \tag{6-10}$$

the IRR% of the simple incremental investment $Y - X$ is the value of $r^*\%$ at which the present worths of the Y and X alternatives are equal. Hence, *either alternative can be chosen.*

When

$$r^*\% > i^*_{Y-X}\%, \tag{6-11}$$

alternative X is the least-cost alternative.

6.7 SUMMARY

Section 6.1 reviews the properties of a simple investment and proves that it has positive intermediate balances.

Section 6.2 shows how to calculate the IRR% of a manufacturing plant investment, and it explains the use of the End-of-Year Convention in specifying the timing of the investment's cash flows. Section 6.3 provides a calculation of the IRR% of a cost reduction investment.

Section 6.4 is a comprehensive analysis of multi-simple investments. It shows that sometimes a multi-simple investment may be a quasi-simple investment. Step-by-Step 6.1 shows how to determine when this is so. Two examples illustrate its use.

Section 6.5 shows how to use *Mathematica* to find the roots of the GPE when MathCad and MATLAB are not available.

Section 6.6 shows that the incremental investment and least-cost criteria predict the same least-cost alternative.

6.8 PROBLEMS

(P6-1) The cash flows of a tri-simple investment are given below. **(Part A)** *Calculate all the roots of the GPE.* **(Part B)** *If there is a unique value of the IRR%, calculate all of the balances.* **(Part C)** *Is this a quasi-simple investment?*

CASH FLOW, $\$10^3$	**−9**	3	3	3	**−2**	3	3	**−5**	3	3	3
YEAR *k*	0	1	2	3	4	5	6	7	8	9	10

(P6-2) *If the fixed capital of the chemical plant investment in Spreadsheet 6.1 is disbursed equally over* ***three*** *years, what is the IRR% of the investment?* NOTES: (1) Everything else remains the same. (2) You may modify Spreadsheet 6.1 to find the IRR%.

(P6-3) At the end of the 6th month of the spec house investment described in Example 3.4, a buyer went into contract to purchase the house. She gave the builder a \$40,000 down payment at the *end* of the 6th month and the remaining \$365,000 at the end of the 11th month. The *modified* monthly cash flows of the spec house are listed in Table 6.2. **(Part A)** *What type of investment is this?* **(Part B)** *Does it have a unique positive root of the General Polynomial Equation? Show how you arrived at your answer.* **(Part C)** *If the answer to Part B is "yes", are the intermediate balances positive?* **(Part D)** If the answer to Part C is "yes", *what is the annual IRR% of this investment?*

TABLE 6.2 **Monthly Cash Flows of the Spec House with a Down Payment**

MONTH	TRANSACTION	CASH FLOW, $
0	Land	–125,000
1	House Plans	–2,500
2	Survey	–1,000
3	Building Permit	–150
4	Foundation	–20,000
5	Framing, Roof, Windows & Siding	–75,000
6	Rough Plumbing & Electrical **+ $40,000**	**19,000**
7	Sheetrock, Tape & Spackle	–16,000
8	Flooring	–22,000
9	Plumbing & Electrical Fixtures	–10,000
10	Kitchen Cabinets, Fixtures & Appliances	–30,000
11	Landscaping & Decking **+ $365,000**	**361,200**

7

Non-Simple Investments

OBJECTIVES

A **non-simple investment** is any investment that is not a simple or quasi-simple investment. Non-simple investments occur less frequently than simple investments, but, as you will see, their cash flow patterns occur in *ordinary* business situations. After reading this chapter, you will be able to calculate the IRRs of the five common *non-simple* investments—***NSI*-Class** problems:

1. Down Payment Investment—***DPI*-Class** problem
2. Termination Investment—***TI*-Class** problem
3. Down Payment & Termination Investment—***DP&TI*-Class** problem
4. Bi-Simple Investment—***BSI*-Class** problem
5. Tri-Simple Investment—***TSI*-Class** problem

7.1 FIVE FUNDAMENTAL PROPERTIES of SIMPLE & QUASI-SIMPLE INVESTMENTS

To help you identify a non-simple investment, let's quickly review the five fundamental properties of simple and quasi-simple investments. ***If an investment fails to have any one of them, it is a non-simple investment.*** *Hence, we must modify our method for calculating the IRR.*

7.1.1 Negative First Cash Flow

It follows from the *definition* of a simple investment that it has a *negative* initial cash flow CF_0. When the value of CF_0 of a *non-simple* investment is *positive*, the initial balance B_0 is *negative*. Hence, ***the investor (lender) owes the investment (borrower) at the beginning of the investment (loan)!***

7.1.2 Positive Last Cash Flow

It follows from the definition of a simple investment that it has a positive last cash flow CF_N. When the value of CF_N of a *non-simple* investment is *negative*, the last intermediate balance B_{N-1} is *negative*. This follows from the balance equation:

$$B_k = B_{k-1}(1+i) - CF_k \qquad k = 1, 2, \ldots, N, \tag{7-1}$$

where B_N is *defined* as zero. Since $(1 + i^*)$ is positive and CF_N is negative, B_{N-1} is *negative*. Hence, ***the investor (lender) owes the investment (borrower) at the end of the (N − 1)th interest period!***

7.1.3 Odd Number of Cash Flow Sign Changes

It follows from the definition of a simple investment that it has one cash flow sign change. If the cash flows of a *non-simple* investment have an *even* number of sign changes, Descartes' Rule of Signs tells us that the number of positive roots of the GPE equals the even number of sign changes or differs from it by an even number. Hence, ***a non-simple investment with an even number of cash flow sign changes can never have one positive root of the GPE and, hence, an IRR!***

7.1.4 Unique Rate of Return

Since a simple investment has only one cash flow sign change, Descartes' Rule of Signs tells us there is only one positive root of the GPE. Hence, a simple investment has a *unique* rate of return (ROR) called the internal rate of return (IRR). However, ***an investment with three or more odd cash flow sign changes may have more than one positive root of the GPE and, hence, multiple RORs!***

7.1.5 Positive Balances

The *initial* and *intermediate* balances of a simple investment are positive. ***If any balance of a non-simple investment is negative, the investor (lender) owes the investment (borrower) at the corresponding timeline point!***

7.1.6 Fundamental Properties of the Four Basic Non-Simple Investments

Table 7.1 lists the **four basic non-simple investments** and any of the five fundamental properties of a simple investment possessed by them. The next four sections explain how to calculate the IRR of each of these non-simple investments.

7.2 DOWN PAYMENT INVESTMENTS

7.2.1 Definition, Properties & Occurrence

The cash flow pattern of a **down payment (DP) investment** is *a simple investment pattern* ***preceded*** *by one or more positive cash flows.* Figure 7.1 is a CFD of a down payment investment. Table 7.1 shows that a down payment investment does not have a *unique* ROR and *positive* balances. However, I will soon *define* a *mixed* down payment investment that does (see Subsection 7.2.4). *Down payment investments usually occur when a purchaser contracts a company to provide a product or service and the company requires a down payment on the purchase price before it begins work.*

TABLE 7.1 Properties of the Four Basic Non-Simple Investments

FUNDAMENTAL PROPERTIES of a SIMPLE INVESTMENT	PROPERTIES of the FOUR BASIC NON-SIMPLE INVESTMENTS			
	DOWN PAYMENT	TERMINATION	DOWN PAYMENT & TERMINATION	MULTI-SIMPLE
NEGATIVE FIRST CASH FLOW	✗	✔	✗	✔
POSITIVE LAST CASH FLOW	✔	✗	✗	✔
ODD NUMBER of CASH FLOW SIGN CHANGES (1)	✗ (2)	✗ (2)	✔ (3)	✔ $(2N-1)^1$
UNIQUE RATE of RETURN	✗	✗	Possible	Possible
POSITIVE INITIAL and INTERMEDIATE BALANCES	✗	✗	✗	Possible

[1] N is the number of simple investment patterns in a multi-simple investment.

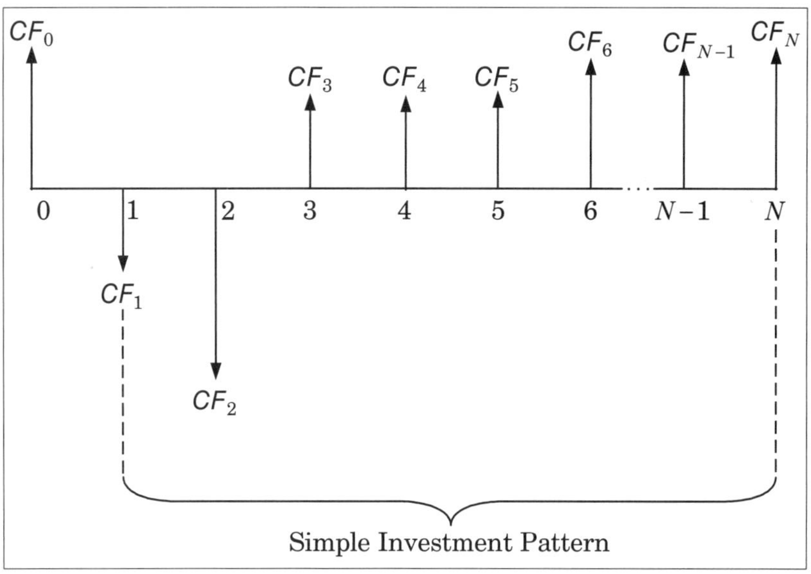

Figure 7.1 **CFD of a Down Payment Investment with One *Positive Initial* Cash Flow**

7.2.2 Finding the RORs of a Down Payment Investment

Since a down payment investment has two sign changes of its cash flows, there may be two or no values of i^* that satisfy the GPE. Let's explore a down payment investment with two values of i^*. Doing so will show us how to calculate a *unique* ROR with *positive* initial and intermediate balances.

EXAMPLE 7.1 Finding the ROR%s and Balances of a Down Payment Investment

Problem: A builder is negotiating with a town to build 20 low income houses that he can sell for $250,000 each. As an *incentive*, the town agrees to pay the builder $\$1 \times 10^6$ when the contract is signed. The estimated cash flows of the project are shown below. *Find the ROR%s and balances of this down payment investment.*

CASH FLOW, $\$10^6$	1	−2.5	−1.5	2.5	1.5	1
YEAR *k*	0	1	2	3	4	5

Solution: It's convenient to use Mathcad, MATLAB, or a graphing calculator to find the roots of the GPE. Here's the Mathcad result:

$$CF := \begin{pmatrix} 1 \\ 1.5 \\ 2.5 \\ -1.5 \\ -2.5 \\ 1 \end{pmatrix} \qquad polyroots(CF) = \begin{pmatrix} -1.0000 \\ -0.2294 + 0.4801i \\ -0.2294 - 0.4801i \\ 1.3589 \\ 2.5998 \end{pmatrix}$$

Subtracting 1 from each positive root and multiplying the results by 100 yields the two ROR%s: 35.89% per year and a whopping 159.98% per year. Let's compute these ROR%s and their balances in Excel. *This will help us figure out how we can obtain a unique ROR% with positive balances.*

The two ROR%s calculated by Excel are shown in Spreadsheet 7.1. As expected, they are identical to those obtained with Mathcad. The initial balance for each interest rate is the *negative* value of the down payment ($-1 \times \$10^6$), and, as required, the *last* balances are zero. I will now explain the meaning of the negative balances in Spreadsheet 7.1. Then I will show you how to find a *unique* ROR% of a down payment investment that has *positive* initial and intermediate balances.

	A	B	C	D	E	F
1						
2		***ROR%s and Balances of a Down Payment Housing Investment***				
3		***k***	CF_k, $\$10^6$	$B_k = B_{k-1}(1+i^*) - CF_k$	CF_k, $\$10^6$	$B_k = B_{k-1}(1+i^*) - CF_k$
4		0	1	(1)	1	(1)
5		1	(2.5)	1.1411	(2.5)	(0.0998)
6		2	(1.5)	3.0506	(1.5)	1.2404
7		3	2.5	1.6454	2.5	0.7249
8		4	1.5	0.7359	1.5	0.3846
9		5	1	0.0000	1	0.0000
10			**35.89%**	**=IRR(C4:C9)**	**159.98%**	**=IRR(F4:F9, 100%)**

Spreadsheet 7.1 NOTE: Excel calculates 35.89% without a guess. To obtain the 159.98% value, we must provide the IRR function with a guess that is close to it.

7.2.3 Meaning of Negative Balances

In the derivation of the GRE and, hence, the GPE, we only *assume* that (1) the interest rate is constant and (2) the last balance B_N is zero. Hence, ***B_0 and one or more intermediate balances of a non-simple investment can be negative.*** It follows from the balance equation

$$B_k = B_{k-1}(1+i^*) - CF_k \tag{7-2}$$

that *a negative balance earns interest at the same interest rate i* earned by the positive balances of the non-simple investment.* But where does a negative balance earn this interest?

> *The balance equation implies that we can deposit negative balances in an **external** investment with the **same interest rate i*** earned by the positive balances.*

For example, when the ROR% in Spreadsheet 7.1 is 159.98% per year, the $\$1 \times \$10^6$ down payment, which is a negative initial balance, is *mathematically* invested in an *external* investment that also earns 159.98% per year. The balance at the end of the first year is given by the balance equation:

$$\begin{aligned} B_1 &= B_0(1+i^*) - CF_1 \\ &= \underbrace{-\$1\times10^6(1+1.5998)}_{\text{Money in an External Investment}} - \underbrace{(-\$2.5\times10^6)}_{\text{Project Disbursement}} \\ &= -\$2.5998\times10^6 + \$2.5\times10^6 \\ &= -\$0.0998\times10^6. \end{aligned}$$

The balance is still negative. Hence, *even though the project has a disbursement of $\$2.5 \times 10^6$, it does not come from the builder!* It is subtracted from the balance of the external investment at the end of the first year.

The small negative balance B_1 remains *mathematically* invested in the *external* investment at 159.98% per year. At the end of the second year, the project balance is $\$1.2404 \times 10^6$. *This is the amount of money that the builder must invest at the end of the second year.* ***This negative cash flow and the remaining positive cash flows of the project constitute a simple investment whose ROR% is 159.98%.*** The following subsection explains how we can transform a down payment investment into a simple investment by means of a *realistic* external investment.

7.2.4 Mixed Down Payment Investment

Definition

In the explanation of negative balances in Subsection 7.2.3, the negative balances of the down payment investment are invested in an *external* investment with the *same* interest rate as the ROR. Is this *mathematical* scenario realistic? Of course not. If the builder has an external investment yielding 159.98% per year or even 35.89% per year, why would he assume the risks associated with the building project?

We can eliminate the ambiguity of multiple RORs and the implausibility of negative balances invested externally at the RORs of a down payment investment by using the **External Investment Assumption**:

External Investment Assumption

The negative balances of a down payment investment are invested in an external investment with a specified interest rate i_E.

Hence, a **mixed down payment investment** is a sequence of an external investment with a *specified* interest rate i_E and a simple internal investment. Step-by-Step 7.1 explains how we calculate the IRR%:

Calculating the IRR% of a Mixed Down Payment Investment: Step–by–Step 7.1

(1) *Specify* the external investment's interest rate i_E and its negative *initial* balance:

$$B_{E,0} = -CF_0 \qquad (7\text{-}3)$$

(2) *Calculate* the balance of the external investment at the end of the first interest period:

$$B_{E,1} = B_{E,0}(1 + i_E) - CF_1. \qquad (7\text{-}4)$$

If $B_{E,1}$ is *positive*, go to Step 3. If $B_{E,1}$ is *negative*, calculate subsequent balances until one is positive. Then go to Step 3.

(3A) *Convert* the *first* positive balance from Step 2 to *the first and possibly only* ***negative*** *cash flow of the* ***simple internal*** *investment.* To signify the transfer to the simple internal investment, you can insert an *optional* arrow that points to the cell with its *first* cash flow: →

(3B) The cash flows *not used* in Step 2 are the *subsequent* cash flows of the simple internal investment.

(4) *Calculate* the IRR% of the simple *internal* investment, which is defined as a **Mixed Down Payment Investment (*MDPI*)**.

Choice of the Specified External Interest Rate

An external investment should have a low risk of default. Typical low-risk external investments are bank certificates of deposit and US Treasury notes. Their interest rates depend on the prevailing economic conditions; values in the range of 3% per year to 6% per year are common.

Some authors use the MARR% (the minimum IRR% that a company will accept when it assumes the *risk* of a typical business investment) for the specified value of the external interest rate. But *this is a risky optimistic choice.* Using the MARR% for the external interest rate will yield an IRR% that is larger than that obtained with a *low risk* external interest rate. However, *this larger IRR% may cloud our judgement about accepting the underlying down payment investment.*

EXAMPLE 7.2 Calculating the IRR% of a Mixed Down Payment Investment (*MDPI*-Class)

Problem: *Calculate the IRR% of the investment in Example 7.1 when negative balances are invested in an external investment with an interest rate of 6% per year.*

Solution: Spreadsheet 7.2 implements Step-by-Step 7.1. It shows that the IRR% of the mixed investment is 28.04% per year.

	A	B	C	D	E	F
1						
2		***IRR% of a Mixed Down Payment Housing Investment***				
3			i_E%	6%		
4				***External Investment***		
5		k	CF_k, \$$10^6$	$B_k = B_{k-1}(1+i_E) - CF_k$	***Simple Investment***	
6		0	1	(1)	CF_S, \$$10^6$	
7		1	(2.5)	1.44 **-D7 →**	(1.44)	**Cash Flows of the *Simple* Internal Investment**
8		2	(1.5)	→	(1.5)	
9		3	2.5	→	2.5	
10		4	1.5	→	1.5	
11		5	1	→	1	
12				=IRR(E7:E11)	**28.04%**	

Spreadsheet 7.2 This spreadsheet implements Step-by-Step 7.1. NOTES: (1) The *initial* balance $B_{E,0}$ of the *external* investment is $-CF_0$. (2) The **Text Box** in cell D7 explains the origin of the first cash flow of the *simple internal* investment, which also has a second *negative* cash flow.

7.2.5 Properties of a Mixed Down Payment Investment

A mixed down payment investment has two important properties:

1. *The IRR of a mixed down payment investment increases with* i_E because it reduces the amount of money that must be invested to receive the positive cash flows of the simple internal investment.
2. *The absence of an ROR of the down payment investment is irrelevant* because it is not used in calculating the IRR of the mixed down payment investment.

Examples 7.3 and 7.4 illustrate these properties.

EXAMPLE 7.3 Effect of Using an MARR% for the External Interest Rate of the Mixed Down Payment Investment (*MDPI*-Class)

Problem: The builder in Example 7.1 uses an MARR% of 15% per year to determine the feasibility of potential projects. *Calculate the IRR% of the mixed down payment investment of Example 7.2 when negative balances are invested in an external investment whose interest rate is the MARR%.*

Solution: Spreadsheet 7.3 calculates the IRR%. Changing i_E% from 6% to 15% per year yields an IRR% of 30.21% per year. This is 7.74% larger than the original IRR% of 28.04% per year. The first disbursement of the internal investment is now $\$1.35 \times 10^6$ instead of the previous value of $\$1.44 \times 10^6$. Thus, *the builder will save \$90,000 if the external investment can earn 15% per year.*

	A	B	C	D		E	F
1							
2		***IRR% of a Mixed Down Payment Housing Investment***					
3			i_E%	15%			
4				***External Investment***			
5		k	CF_k, \$$10^6$	$B_k = B_{k-1}(1+i_E)-CF_k$		***Simple Investment***	
6		0	1	(1)		CF_S, \$$10^6$	
7		1	(2.5)	1.35	**–D7** →	(1.35)	**Cash Flows of the *Simple* Internal Investment**
8		2	(1.5)	→		(1.5)	
9		3	2.5	→		2.5	
10		4	1.5	→		1.5	
11		5	1	→		1	
12					=IRR(E7:E11)	**30.21%**	

Spreadsheet 7.3 Increasing the value of i_E% from 6% to the builder's MARR% of 15% increases the IRR% from 28.04% to 30.21%.

EXAMPLE 7.4 Calculating the IRR% of a Mixed Down Payment Investment When the GPE Has No Positive Root (*MDPI*-Class)

Problem: The architect that designed the low income houses in Example 7.2 noticed that the order of the negative cash flows should be reversed:

CASH FLOW, \$$10^6$	1	–1.5	–2.5	2.5	1.5	1
YEAR k	0	1	2	3	4	5

How does this change affect the IRR% of the mixed down payment investment if the external investment earns 6% per year?

Solution: As with any mixed down payment investment, we calculate its IRR% with Step-by-Step 7.1. But before we do this, let's see how the change in the order of the negative cash flows affects the roots of the GPE. As in Example 7.1, we can use Mathcad, MATLAB, or a graphing calculator to conveniently find the roots. Here's the Mathcad result:

$$CF := \begin{pmatrix} 1 \\ 1.5 \\ 2.5 \\ -2.5 \\ -1.5 \\ 1 \end{pmatrix} \qquad polyroots(CF) = \begin{pmatrix} -1.3667 \\ -0.2505 + 0.4406i \\ -0.2205 - 0.4406i \\ 1.6839 + 0.1141i \\ 1.6389 - 0.1141i \end{pmatrix}$$

Now there are no positive roots of the GPE! And even if there were two, neither would be relevant to calculating the IRR% of the mixed down payment investment.

Spreadsheet 7.4 calculates the IRR% of this mixed down payment investment.

	A	B	C	D		E	F
1							
2		*IRR% of a Mixed Down Payment Housing Investment*					
3			i_E%	6%			
4				*External Investment*			
5		k	CF_k, $10^6	$B_k = B_{k-1}(1+i_E)-CF_k$		*Simple Investment*	
6		0	1	(1)		CF_S, $10^6	
7		1	(1.5)	0.44	–D7 ➡	(0.44)	Cash Flows of the *Simple* Internal Investment
8		2	(2.5)	➡		(2.5)	
9		3	2.5	➡		2.5	
10		4	1.5	➡		1.5	
11		5	1	➡		1	
12				=IRR(E7:E11)		**34.68%**	

Spreadsheet 7.4 Changing the order of the two negative cash flows of the investment increases the IRR% from 28.04% to 34.68. NOTES: (1) Here and in Spreadsheet 7.2, the builder's total disbursement is $2.94 × 10^6, *but now most of it occurs at the end of the first year of the internal investment.* (2) Changing the order of the negative cash flows yields an IRR% that is 23.7% larger than the original IRR% of 28.04% per year. This again illustrates *the time value of money:* ***late receipts (disbursements) are worth less than the same receipts (disbursements) made early.***

7.3 TERMINATION INVESTMENTS

7.3.1 Definition, Properties & Occurrence

The cash flow pattern of a **termination (T) investment** is *a simple investment pattern followed by one or more negative cash flows.* Figure 7.2 is a CFD of termination investment. Table 7.1 shows that a termination investment does not

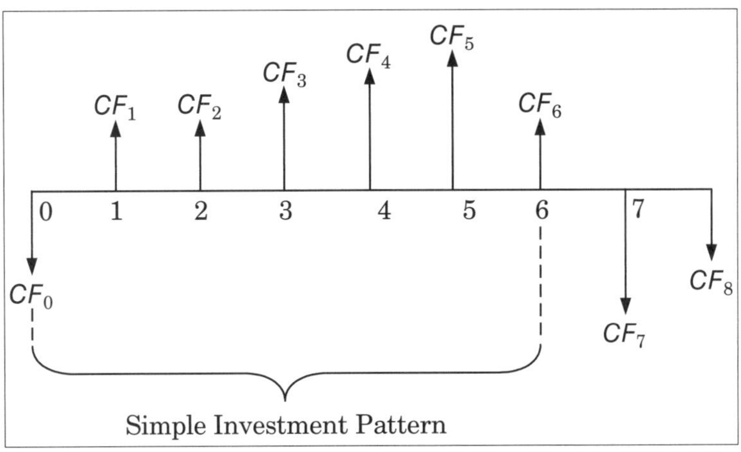

Figure 7.2 **CFD of a Termination Investment with Two *Negative Final* Cash Flows**

have a *unique* ROR and *positive intermediate* balances. However, I will *define* a *mixed* termination investment that does. *Termination investments often occur when an investor leases the mineral rights to a property and must spend money at the end of the lease to restore the property to its natural state.*

7.3.2 Mixed Termination Investment

External Investment with a 0% Interest Rate

To understand a **mixed termination investment** consider the following termination investment:

CASH FLOW, $\$10^6$	−2	0.5	0.5	0.5	1.5	0.9	0.25	−0.75
YEAR *k*	0	1	2	3	4	5	6	7

If we take $\$0.5\times10^6$ from CF_5 and all of CF_6 and put them in an *external* investment with an interest rate of 0%, we would could withdraw $\$0.75\times10^6$ from the external investment to provide the negative cash flow CF_7. The IRR% of this *mixed* termination investment is calculated in Spreadsheet 7.5.

	A	B	C	D
1				
2		*k*	CF_k, $\$10^6$	
3		0	(2)	
4		1	0.5	
5		2	0.5	
6		3	0.5	
7		4	1.5	
8		5	0.4	
9			**18.71%**	**=IRR(C3:C8)**

Spreadsheet 7.5 IRR% of the mixed termination investment when the interest rate of the *external* investment is 0%.

External Investment with a Positive Interest Rate

The interest rate of the external investment is *always* positive; hence, we use Step-by-Step 7.2 to find the IRR of a mixed termination investment:

Calculating the IRR of a Mixed Termination Investment: Step-by-Step 7.2

(1) *Specify* the interest rate of the *external* investment i_E and its final balance:

$$B_{E,N} \equiv 0. \qquad (7\text{-}5)$$

(2) *Calculate* $B_{E,N-1}$ of the *external* investment from a *rearranged* form of the balance equation:

$$B_{E,N-1} = \frac{B_{E,N} + CF_N}{(1+i_E)}, \qquad (7\text{-}6)$$

where $B_{E,N-1}$ is *always* negative because CF_N is always negative.

(3) *Calculate* the preceding balances of the external investment with

$$B_{E,k-1} = \frac{B_{E,k} + CF_k}{(1 + i_E)}, \tag{7-7}$$

until a balance is *positive*.

(4) ***Add*** the *last negative* external balance and its corresponding cash flow. Make this sum the *last positive* cash flow of the *simple internal* investment.

(5) *Calculate* the IRR% of the simple *internal* investment, which is defined as a **Mixed Termination Investment (*MTI*)**.

EXAMPLE 7.5 Calculating the IRR% of a Mixed Termination Investment (*MTI*-Class)

Problem: Calculate the IRR% of the mixed termination investment when the interest rates of the external investment are 0% and 5% per year.

Solution: It's easy to implement Step-by-Step 7.2 in Excel, and Spreadsheet 7.6 shows the result when $i_E\% = 0\%$. As expected it is 18.71%, the result in Spreadsheet 7.5.

	A	B	C	D		E	F
1							
2		***IRR% of a Mixed Termination Investment***					
3						***Simple Internal Investment***	
4		***k***	CF_k, \$10⁶			$CF_{S,k}$, \$10⁶	
5		0	(2)	→		(2)	**Cash Flows of the *Simple* Internal Investment**
6		1	0.5	→		0.5	
7		2	0.5	→		0.5	
8		3	0.5	→		0.5	
9		4	1.5	→		1.5	
10		5	0.9	**(0.5)**	**C10 + D10** →	0.4	
11		6	0.25	(0.75)		**18.71%**	**=IRR(E5:E10)**
12		7	(0.75)	0			
13			$i_E\%$	$B_{E,k-1} = \dfrac{B_{E,k} + CF_k}{1+i}$			
14			0%	***External Investment***			

Spreadsheet 7.6 Calculation of the IRR% of the mixed termination investment. NOTE: The external balance in cell D9 is *positive*. Hence, it is not used.

As expected, the IRR% (18.71%) is identical to that obtained in Spreadsheet 7.5. Spreadsheet 7.7 shows the result when $i_E\% = 5\%$.

	A	B	C	D	E	F
1						
2		***IRR% of a Mixed Termination Investment***				
3					***Simple Internal Investment***	
4		***k***	CF_k, \$$10^6$		$CF_{S,k}$, \$$10^6$	
5		0	(2)	→	(2)	**Cash Flows of the *Simple* Internal Investment**
6		1	0.5	→	0.5	
7		2	0.5	→	0.5	
8		3	0.5	→	0.5	
9		4	1.5	→	1.5	
10		5	0.9	**(0.44)** **C10 + D10** →	0.46	
11		6	0.25	(0.71)	**19.20%**	**=IRR(E5:E10)**
12		7	(0.75)	0		
13			i_E%	$B_{E,k-1} = \dfrac{B_{E,k} + CF_k}{1+i}$		
14			5%	***External Investment***		

Spreadsheet 7.7 Increasing the value of i_E% to 5% yields an IRR% of 19.20, which is a modest 2.62% increase in the value of the IRR%.

7.4 DOWN PAYMENT & TERMINATION INVESTMENTS

7.4.1 Definition, Properties & Occurrence

The cash flow pattern of a **down payment & termination (DP&T) investment** *is a simple investment pattern preceded by one or more positive cash flows and followed by one or more negative cash flows.* Figure 7.3 is a CFD of a DP&T investment.

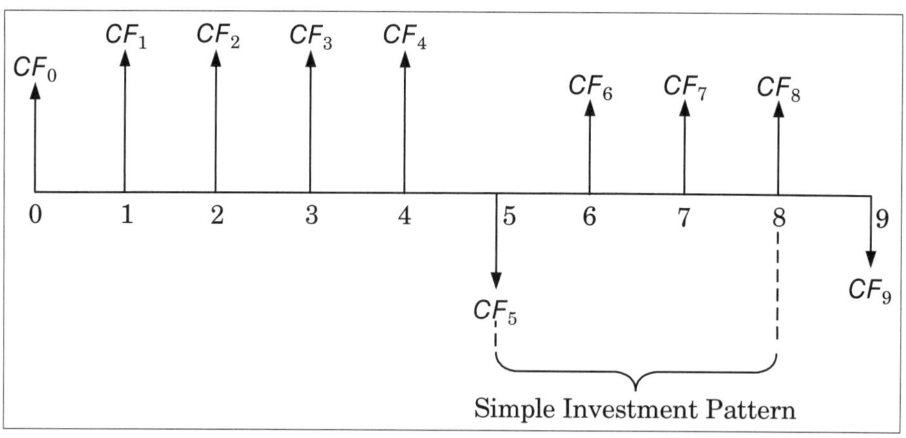

Figure 7.3 **CFD of a DP&T Investment with Five *Positive Initial* Cash Flows and One *Negative Final* Cash Flow**

Table 7.1 shows that a DP&T investment *may* have a unique IRR% because it has three cash flow sign changes, but it never has *positive* balances. However, *a mixed DP&T investment has both properties.*

A DP&T investment often occurs when an oil field that is currently providing positive annual cash flows is leased by its owner, the lessor, to a lessee subject to the following conditions:

1. The lessee pays the lessor a royalty on each barrel of oil produced, but the oil field provides an immediate *positive cash flow series* for the lessee.
2. When the daily production of the oil field becomes uneconomical, the lessee must stop production and inject pressurized water into one or more wells. This pressurizes the oil below the surrounding wells so that they resume economical production. *The lessee pays for this **secondary recovery** project.*
3. When the oil field can no longer economically produce oil after water injection, it is shut down, and the land is restored to its natural state. *The lessee also pays for this.*

7.4.2 Mixed Down Payment & Termination Investment

Definition

A **mixed DP&T investment** is a sequence of (1) an *initial* external investment with a specified interest rate i_{E_1}, (2) a simple *internal* investment, and (3) a *final external* investment with a specified interest rate i_{E_2}.

Calculating the IRR%

We use Step-by-Steps 7.1 and 7.2 to calculate the first and last cash flows of a mixed DP&T investment:

(1) Step-by-Step 7.1

The *beginning* positive cash flows of a mixed DP&T investment are invested in the ***first external*** investment, with an interest rate i_{E_1}, until its balance becomes positive. ***The first positive external balance is the first and possibly only negative cash flow of the simple internal investment.***

(2) Step-by-Step 7.2

The *ending* negative cash flows of a mixed DP&T investment are invested in the ***second external*** investment, with an interest rate i_{E_2}, until its balance becomes positive. ***The sum of the last negative external balance and the corresponding cash flow is the last positive cash flow of the simple internal investment.***

(3) Calculate the IRR% of the Simple Internal Investment

The *first* and *last* cash flows of the simple internal investment and the cash flows *between* them constitute the simple internal investment. Calculate the IRR% of the simple *internal* investment, which is defined as a **Mixed Down Payment & Termination Investment (*MDP&TI*)**.

Computation Note: *The values of the two external investment interest rates,* i_{E_1} *and* i_{E_2}, *can be different.*

EXAMPLE 7.6 Calculating the IRR% of a Mixed Down Payment & Termination Investment (*MDP&TI*-Class)

Problem: The estimated cash flows of an oil field lease are shown below. *Calculate the IRR% when i_{E_1}% and i_{E_2}% are 5% and 4% per year, respectively.*

CASH FLOW, $10^6	3.5	4	3	1.5	−23	5	4	3	2.5	2	1.5	−1.5
YEAR *k*	0	1	2	3	4	5	6	7	8	9	10	11

Solution: We use Step-by-Step 7.1 to find the *first* cash flow of the simple internal investment and Step-by-Step 7.2 to find its *last* cash flow. Then we calculate the IRR% of the resulting *simple* internal investment. All of this is done easily and quickly in Spreadsheet 7.8. It shows that the IRR% of the mixed DP&T investment is 28.47% per year.

	A	B	C	D	E	F
1						
2		***IRR% of a Mixed DP&T Oil Field Lease Investment***				
3			i_{E_1}%	5%		
4				***First External Investment***		
5		*k*	CF_k, \$10^6	$B_k = B_{k-1}(1+i_{E_1}) - CF_k$		
6		0	3.5	(3.5)		
7		1	4	(7.68)		
8		2	3	(11.06)		
9		3	1.5	(13.11)	***Simple Internal Investment***	
10		4	(23)	9.23 **−D10** →	(9.23)	
11		5	5	→	5	
12		6	4	→	4	**Cash Flows of the Simple Internal Investment**
13		7	3	→	3	
14		8	2.5	→	2.5	
15		9	2	→	2	
16		10	1.5	**(1.44)** **C16 + D16** →	0.06	
17		11	(1.50)	0	**28.47%**	**=IRR(E10:16)**
18			i_{E_2}%	$B_{k-1} = \dfrac{B_k + CF_k}{1 + i_{E_2}}$		
19			4%	***Second External Investment***		

Spreadsheet 7.8 The Text Boxes in cells D10 and D16 explain the calculations of the first and last cash flows, respectively, of the internal investment.

7.5 BI-SIMPLE & TRI-SIMPLE INVESTMENTS

7.5.1 Definitions, Properties & Occurrences

The cash flow pattern of a **multi-simple investment** is *a sequence of two or more simple investment patterns. Bi-simple and to a lesser extent tri-simple investments are the multi-simple investments of practical significance.* Figures 7.4 and 6.4 are the CFDs of a bi-simple and tri-simple investment, respectively.

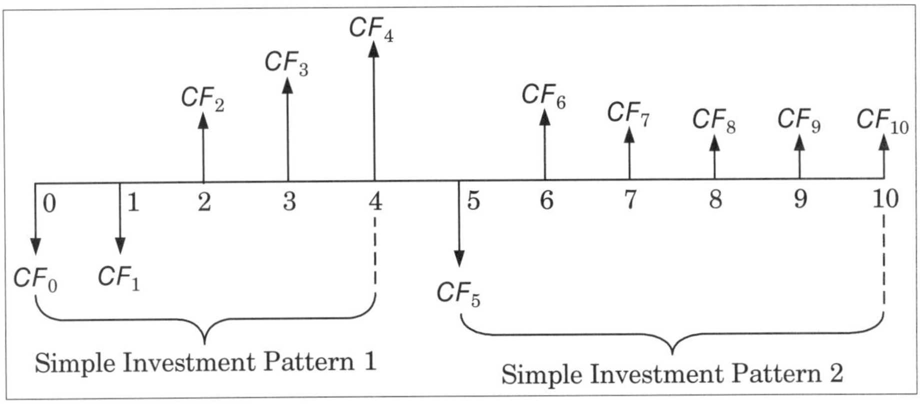

Figure 7.4 **CFD of a Bi-Simple Investment**

A bi-simple investment often occurs when a simple investment has an *unexpected* negative cash flow where a positive one had been predicted. And when a spec house is sold in the *middle* of the construction phase, the normal sequence of negative cash flows is interrupted by the positive down payment. It is followed by another sequence of negative cash flows and the positive cash flow when the construction is completed.

A tri-simple investment ***occasionally*** *occurs when we want to calculate the IRR of the* ***incremental investment*** *of* ***two mutually exclusive investments (MEIs)*** *(see Subsection 8.1.1).*

The cash flows of a multi-simple investment with N simple investment patterns have an **odd number of sign changes** δ_N:

$$\delta_N = 2N - 1. \quad (7\text{-}8)$$

Hence, it follows from Descartes' Rule of Signs that a bi-simple investment may have *one* or three positive roots of the GPE, and a tri-simple investment may have *one,* three, or five.

Bi-simple and tri-simple investments are often *quasi-simple* investments. That is, they have *unique positive roots* of the GPE and *positive intermediate balances.* We use Step-by-Step 6.1 in Subsection 6.4.2 to determine if a bi-simple or tri-simple investment is a quasi-simple investment.

7.5.2 Mixed Bi-Simple Investment

Definition

Here are the cash flows of the bi-simple tractor factory investment in Example 6.1:

CASH FLOW, $\$10^6$	–15	–15	15	20	25	**–10**	15	10	5	5	5	5
YEAR k	0	1	2	3	4	5	6	7	8	9	10	11

It is shown there that the value of the *unique* $i^*\%$ is 35.77% and that the intermediate balances are *positive*. However, as the *magnitude* of the negative cash flow of the *second* simple investment pattern increases, one negative balance or several *adjacent* ones are generated. For example, if CF_5 *decreases* from $-\$10 \times 10^6$ to $-\$30 \times 10^6$, a polynomial root finder returns a *unique* ROR% of 27.58% per year. But Spreadsheet 7.9 shows that B_4 is now *negative*:

	A	B	C	D
1				
2		***Effectof Decreasing CF*** $_5$ ***on the ROR% and Balances of the Bi-Simple Tractor Factory Investment***		
3		k	CF_k, $\$10^6$	$B_k = B_{k-1}(1+i^*) - CF_k$
4		0	(15)	15
5		1	(15)	34.1365
6		2	15	28.5502
7		3	20	16.4235
8		4	25	**(4.0475)**
9		5	**(30)**	24.8364
10		6	15	16.6854
11		7	10	11.2867
12		8	5	9.3992
13		9	5	6.9913
14		10	5	3.9192
15		11	5	0.0000
16			**27.58%**	**=IRR(C4:C15)**

Spreadsheet 7.9 The *increase* in the *magnitude* of CF_5 from 10 to 30 generates the *negative* balance B_4.

Since a negative balance must be invested in an external investment, a **mixed bi-simple investment** is an *external* investment *sandwiched* between *two simple internal investments* with the *same* IRR.

Calculating the IRR%

We use Step-by-Step 7.3 to calculate the IRR% of a mixed bi-simple investment.

Calculating the IRR% of a Mixed Bi-Simple Investment in Excel: Step-by-Step 7.3
(1) Provide a *guess* for the IRR% ($i^*\%$) that is the ROR% calculated from the cash flows of the bi-simple investment. NOTE: Usually only *one* ROR% exists. If there are two values, use the smaller one.

(2) Specify the *external* investment's percentage interest rate: $i_E\%$.

(3) Calculate the balances of the ***first simple internal investment*** with the balance equation

$$B_k = B_{k-1}(1 + i^*) - CF_k \tag{7-9}$$

until a balance is *negative*. *If a negative intermediate balance does not occur, reduce the value of i*%.*

(4) Make the *negative* balance from Step 3 the *first* balance of the ***external investment.***

(5) Calculate the ***external investment*** balances with the balance equation

$$B_{E,k} = B_{E,k-1}(1 + i_E) - CF_k \tag{7-10}$$

until a balance is *positive*.

(6) Make the *positive* balance from Step 5 the *first* balance *of the* ***second simple internal investment.***

(7) Calculate the balances of the ***second simple internal investment*** with Equation (7-9). If the last balance $\boldsymbol{B_N}$ is not zero, go to Step 8.

(8) Use the **Goal Seek** command to make the value of $\boldsymbol{B_N} = \mathbf{0.00}$ by changing the value of $i^*\%$. To do this,

(8.1) *Click* the **Data** tab. In the **Data Tools** group *Click* **What-If Analysis** > *Click* **Goal Seek...** to open the **Goal Seek** dialog box.

(8.2) In the **Set cell:** text box *Click* the cell that contains $\boldsymbol{B_N}$ (the value you want to change).

(8.3) In the **To value:** text box *Type* **0.00** (the desired value of $\boldsymbol{B_N}$).

(8.4) In the **By changing cell:** text box *Click* the cell that contains ***i*%*** (the value you want Excel to change) > *Click* **OK**.

(8.5) After the iteration converges, *Click* **OK** in the **Goal Seek** status box.

EXAMPLE 7.7 Calculating the IRR% of a Mixed Bi-Simple Investment (*MBSI*-Class)

Problem: *Use the cash flows in Spreadsheet 7.9 to find the IRR% of the mixed bi-simple tractor factory investment when the interest rate of the external investment is 5% per year.*

Solution: Spreadsheet 7.10 is the implementation of Steps 1 through 7 of Step-by-Step 7.3. It shows the spreadsheet after the *first* iteration. As you can see, the value of B_N is 3.97. However, we can easily find the value of ***i*%*** that makes $\boldsymbol{B_N} = \mathbf{0}$ by using Goal Seek to iterate according to Steps 8.1 through 8.5 in Step-by-Step 7.3. Goal Seek iterates by changing the value of ***i*%*** in value cell D3 until the value of $\boldsymbol{B_N}$ in formula cell D17 equals **0.00**.

Spreadsheet 7.11 shows that this occurs when the IRR% is **27.09%** per year. As expected, this is lower than the value of 27.58% obtained in Spreadsheet 7.9. In that calculation the small negative balance is *mathematically* invested in an external investment with the same interest rate as the simple internal investments: 27.58% per year. In fact, *if we define the external interest rate as 27.58%, Goal Seek returns a value of 27.58% for the converged value of* ***i*%***.

	A	B	C	D		E		F
1								
2				***IRR% of the Mixed Bi-Simple Tractor Factory Investment***		***First Iteration***		
3			$i^*\%$	**27.58%**		5%		$i_E\%$
4				***First Internal Investment***				
5		k	CF_k, \$$10^6$	$B_k = B_{k-1}(1+i^*) - CF_k$				
6		0	(15)	15				
7		1	(15)	34.14				
8		2	15	28.55		***External Investment***		
9		3	20	16.43		$B_{E,k} = B_{E,k-1}(1+i_E) - CF_k$		
10		4	25	**(4.04)**	**D10 ⟶**	(4.04)		
11		5	**(30)**	25.75		**⟵ E11**	**25.75**	
12		6	15	17.86		**Balances of the 2nd Simple Internal Investment**		
13		7	10	12.78				
14		8	5	11.31				
15		9	5	9.43				
16		10	5	7.03				
17		11	5	**3.97**				

Spreadsheet 7.10 After the first iteration, the value of $B_N = B_{11} = 3.97$. However, Spreadsheet 7.11 shows that when Goal Seek converges, $i^*\% = 27.09\%$, and $B_N = B_{11} = 0.00$.

Confirmation of the Definition of a Mixed Bi-Simple Investment

Spreadsheet 7.11 also confirms that *the IRR%s of the two simple internal investments are both 27.09%.*

7.5.3 Mixed Tri-Simple Investment: Two Simple Internal Investments

Definitions

As the *magnitude* of a negative cash flow of the *second* or *third* simple investment pattern of a tri-simple investment (see Figure 6.4) increases, *one* negative balance or several *adjacent* ones are generated. Hence, a **mixed tri-simple investment** *usually* has an *external* investment *sandwiched* between *two simple internal investments* with the same IRR. The less common case of *two separate* negative balances is not treated here.

Calculating the IRR

When a mixed tri-simple investment has *one* negative balance or several adjacent ones, we use Step-by-Step 7.3 to calculate the *common* IRR of the *two* simple internal investments.

	A	B	C	D	E	F
1						
2				***IRR% of the Mixed Bi-Simple Tractor Factory Investment***	***Converged***	
3			$i^*\%$	**27.09%**	5%	$i_E\%$
4				***First Internal Investment***		
5		k	CF_k, \$10^6	$B_k = B_{k-1}(1+i^*) - CF_k$		
6		0	(15)	15		
7		1	(15)	34.06		
8		2	15	28.29	***External Investment***	
9		3	20	15.96	$B_{E,k} = B_{E,k-1}(1+i_E) - CF_k$	
10		4	25	**(4.72)** **D10 →**	(4.72)	
11		5	**(30)**	25.04	**← E11** **25.04**	
12		6	15	16.82	**Balances of the Second Simple Internal Investment**	
13		7	10	11.38		
14		8	5	9.47		
15		9	5	7.03		
16		10	5	3.93		
17		11	5	**0.00**		
18				***Confirmation of the Two Internal IRRs***		
19				**Internal Investment 1**	**Internal Investment 2**	
20				(15)	(25.04)	
21				(15)	15	
22				15	10	
23				20	5	
24				20.28	5	
25	**=IRR(D20:D24)**			**27.09%**	5	
26					5	
27				**=IRR(E20:E26)**	**27.09%**	

Spreadsheet 7.11 Converged solution of the mixed bi-simple investment and confirmation of the IRR%.

7.6 SUMMARY

Section 7.1 reviews the five fundamental properties of a simple investment and Table 7.1 identifies those that are not possessed by the four classes of non-simple investments. Section 7.2 explains the meaning of negative balances in non-simple investments and introduces the External Investment Assumption: *negative balances are invested in an external investment with a specified interest rate* i_E. Subsection 7.2.4 explains how to calculate the IRR% of a mixed down payment investment with Step-by-Step 7.1, and it is used in Subsection 7.2.5 to illustrate the properties of mixed down payment investments.

Section 7.3 defines a termination investment and explains how to calculate the IRR% of a mixed termination investment with Step-by-Step 7.2. Examples 7.5 and 7.6 illustrates its use. Section 7.4 defines a down payment & termination (DP&T) investment and explains how to use Step-by-Steps 7.1 and 7.2 to calculate the IRR of a mixed DP&T investment. Example 7.6 illustrate the procedure.

Subsection 7.5.1 defines bi-simple and tri-simple investments, and Subsection 7.5.2 explains how to calculate the IRR of a mixed bi-simple investment with Step-by-Step 7.3. Example 7.7 illustrates the procedure. Subsection 7.5.3 explains that the IRR of a mixed tri-simple investment with two simple internal investments can also be calculated with Step-by-Step 7.3.

7.7 PROBLEMS

(P7-1) A yacht builder built a custom yacht for a customer, and the project had the following cash flows:

CASH FLOW, $\$10^6$	0.4	−1.5	−2.5	4.75
YEAR *k*	0	1	2	3

What IRR% did yacht builder earn if the external investment was 4% per year?

(P7-2) The *actual* cash flows of a chemical plant that produced a herbicide are shown below:

CASH FLOW, $\$10^6$	−50	20	20	20	20	14	−10	−5
YEAR *k*	0	1	2	3	4	5	6	7

The receipt stream in the *proposal* for this investment was a uniform series of ten $\$20 \times 10^6$ receipts that yielded an *anticipated* IRR% of 38.45% per year. However, a competitor entered the herbicide market at the end of the 4th year, and the receipt for the 5th year was only $\$14 \times 10^6$. In addition, after the 5th year, the EPA shut the plant down because it was polluting an aquifer that provides the drinking water of a nearby town. The plant site required extensive decontamination at a cost of $\$15 \times 10^6$, of which $\$10 \times 10^6$ was spent in the 6th year.
What was the IRR% of this investment if an external investment of 7% was available?

(P7-3) A mineral lease had the following cash flows:

CASH FLOW, $\$10^6$	3	3	3	2.5	−5	4	4	−6
YEAR *k*	0	1	2	3	4	5	6	

What was the IRR% of this lease if the two external investments were both 6% per year?

8

Mutually Exclusive Investments

OBJECTIVES

In a least-cost study we compare the present or annual worths of the cash flows of two or more mutually exclusive cost alternatives. *Choosing the least-cost alternative precludes selecting additional cost alternatives.* A similar situation occurs when we compare **mutually exclusive investments MEI**s.

After reading this chapter, you will be able to

1. Perform investment studies to find the best simple investment when there are *two* of them, ***MEI*-Class** problems.
2. Perform investment studies to find the best simple investment when there are *three or more* of them, ***MEI*-Class** problems.

8.1 COMPARING TWO SIMPLE MUTUALLY EXCLUSIVE INVESTMENTS

8.1.1 Definitions

Simple **mutually exclusive investments** (MEIs) are simple investments that have the *same* lives and satisfy the *same* need; *choosing the best one precludes selecting additional ones.* For example, a company wants to construct an office building on land that it owns, so that it can receive rental income for at least 10 years. It is considering five different building designs that satisfy this need. The *estimated* construction costs, operating costs, and rental incomes of the buildings are different because they have different footprints, heights, and, hence, office areas. The five building investments are mutually exclusive investments because only one building can be built on the site. Here are two other examples of mutually exclusive investments:

- A chemical can often be produced at a *specified* plant site by two or more competing processes that have *different* fixed and working capitals and annual cash flows.
- Manufacturing plants often invest in new heating equipment to reduce their fuel costs. The cost reductions normally increase with the capital required for different oil or gas fired furnaces.

Two or more investments are **independent** when the selection of *any* one of them does not preclude a decision to accept other investments in the set of independent investments. For example, in 1974 the Mobil Oil Corporation introduced Mobil 1, the first synthetic motor oil. At the same time they were exploring the purchase of the Montgomery Ward department store chain, and they did so in 1976. Building a Mobil 1 production facility and purchasing Montgomery Ward are two interesting independent investments funded by the Mobi-Oil Corporation in the mid-1970s.

Often, capital may not be available to fund two or more attractive independent investments. The strategy for choosing independent investments when capital is limited is called **capital budgeting**. It is carried out at the highest levels of corporate management and is beyond the scope of this book.

8.1.2 Basic MEI Selection Model

Two MEIs

To understand the procedure for comparing *two* MEIs, let's consider two proposed mutually exclusive cost reduction investments *X* and *Y*. Their economic data is presented in Table 8.1. NOTE: Salvage values are negligible.

TABLE 8.1 Economic Data of Two Mutually Exclusive Cost Reduction Investments[1]

ECONOMIC VARIABLE	ESTIMATED VALUES	
	MEI *X*	MEI *Y*
Capital *C*	$100,000	$130,000
Annual Cash Flow *A*	$27,000	$32,500
Life *L*	10 a	10 a
IRR% *i**%	23.8% per year	21.4% per year
MARR% *r**%	20% per year	

[1] As usual, the number of *annual* interest periods is given by $N = \frac{L}{p} = \frac{10\text{a}}{1\text{a}} = 10$.

Computation Note: *We can quickly calculate the IRR% of any investment with* *<u>one</u> capital C and a uniform series of N cash flows A with Excel's* ***RATE*** *function:*

$$\text{RATE}(N, A, -C). \tag{8-1}$$

Thus, for MEI *X*, RATE(10, $27,000, −$100,000) evaluates to 23.8% per year.

Two Assumptions

We make two assumptions to compare *two* mutually exclusive investments *X* and *Y*:

1. Capital is available to fund MEI *X* <u>or</u> MEI *Y*, where the total capital TC_Y required for MEI *Y* is *greater* than that of MEI *X*.
2. Unlike a least-cost comparison, **doing nothing** is an alternative to MEI *X and* MEI *Y* when their IRRs are less than the MARR.

Incremental IRR Selection Criterion

Table 8.1 shows that the IRR%s of MEIs *X and Y exceed* the MARR% of 20% per year. To select the correct MEI, we define the **incremental investment of MEI *Y* versus**

MEI *X*, which is denoted by $Y-X$, as the difference of their cash flows:

$$CF_{Y-X} \equiv CF_Y - CF_X. \tag{8-2}$$

Then we determine its IRR i^*_{Y-X} and use the following selection criteria:

Incremental IRR Selection Criteria
When $i^*_X \geq r^*$ and $i^*_Y \geq r^*$, calculate $i^*_{Y\text{-}X}$. If $i^*_{Y\text{-}X} \geq r^*$, **select MEI *Y***, otherwise **select MEI *X***.

In other words, ***we select MEI Y if the extra capital invested in it also yields an IRR that exceeds or equals the MARR.***

Table 8.2 shows that the IRR% of the incremental investment of the two mutually exclusive cost reduction investments is only 12.9% per year. Since this is less than the MARR% of 20% per year, *we select investment X.*

TABLE 8.2 Expanded Data of Two Cost Reduction MEIs

ECONOMIC VARIABLE	**ESTIMATED VALUES**		
	MEI *X*	**MEI *Y***	**INCREMENTAL INVESTMENT *Y* − *X***
Capital C	\$100,000	\$130,000	\$30,000
Annual Cash Flow A	\$27,000	\$32,500	\$5,500
Life L	10 a	10 a	10 a
IRR% i^*%	23.8% per year	21.4% per year	12.9% per year
MARR% r^*%	20% per year		

Basic MEI Selection Procedure

We use Step-by-Step 8.1 to compare MEIs *X* and *Y*:

Basic MEI Selection Procedure: **Step-by-Step 8.1**
(1) Arrange the cash flows of MEIs *X* and *Y* in order of *increasing* total capitals and calculate the IRRs of the MEIs with Excel's IRR function. **(2)** If $i^*_X < r^*$ and $i^*_Y < r^*$, **do nothing**. **(3A)** If $i^*_X \geq r^*$ and $i^*_Y < r^*$, **select MEI *X***. **(3B)** If $i^*_X < r^*$ and $i^*_Y \geq r^*$, **select MEI *Y***. **(4A)** If $i^*_X \geq r^*$ and $i^*_Y \geq r^*$, that is, the IRR of *each* MEI equals or exceeds the MARR, calculate $i^*_{Y\text{-}X}$. **(4B)** If $i^*_{Y\text{-}X} \geq r^*$, **select MEI *Y***, otherwise **select MEI *X***.

The two assumptions and the Basic MEI Selection Procedure constitute the **Basic MEI Selection Model**.

EXAMPLE 8.1 Finding the Best Rental House Investment (*MEI*-Class)

Problem: A builder wants to build a rental house on a riverfront lot in an Atlanta suburb. He is considering a basic house MEI X and an upscale house MEI Y. Their *predicted* cash flows over an *assumed* 10-year life and their IRR%s are shown in columns D and E of Spreadsheet 8.1. NOTE: The builder *assumes* that in 10 years the houses can be sold for 150% of their cost. *If the builder's MARR% is 15% per year, which house should he build?*

Solution: Spreadsheet 8.1 also shows the cash flows of the incremental investment Y–X and its IRR%. *The builder should build the Y house because its IRR% and the IRR% of the incremental investment are both greater than the MARR%.*

	A	B	C	D	E	F
1						
2		***IRR%s and Incremental IRR% of Two Rental Houses***				
3			***k***	CF_X, \$	CF_Y, \$	CF_{Y-X}, \$
4			0	(290,000)	(400,000)	(110,000)
5			1	36,000	55,000	19,000
6			2	36,150	55,500	19,350
7			3	36,300	56,000	19,700
8			4	36,450	56,500	20,050
9			5	36,600	57,000	20,400
10			6	36,750	57,500	20,750
11			7	36,900	58,000	21,100
12			8	37,050	58,500	21,450
13			9	37,200	59,000	21,800
14			10	472,350	659,500	187,150
15				**15.05%**	**16.45%**	**20.16%**
16				**=IRR(D4:D14)**	**=IRR(E4:E14)**	**=IRR(F4:F14)**

Spreadsheet 8.1 Cash flows and annual IRR%s of MEIs X and Y, and $Y-X$.

8.1.3 Relationships Between the Three IRRs

When the incremental investment $Y-X$ is a *simple* investment, the values of i_Y^*, i_X^*, and i_{Y-X}^* are related by the **Incremental IRR Relations**:

$$\text{If } i_Y^* > i_X^*,\ i_{Y-X}^* > i_Y^*. \tag{8-3}$$

$$\text{If } i_Y^* = i_X^*,\ i_{Y-X}^* = i_Y^*. \tag{8-4}$$

$$\text{If } i_Y^* < i_X^*,\ i_{Y-X}^* < i_Y^*. \tag{8-5}$$

Figures 8.1 and 8.2 show why this is so. The present worth function of any *simple* investment $PW_S(i^*)$ *decreases* monotonically from a value $PW_S(0)$, the sum of its cash flows, to a limiting value $CF_{S,0}$, its *initial* cash flow, as $i^* \rightarrow \infty$.

Hence, i^*_{Y-X} *is the* ***unique*** *intersection point of the* $PW_Y(i^*)$ *and* $PW_X(i^*)$ *curves.* The curves in Figures 8.1 and 8.2 illustrate this.

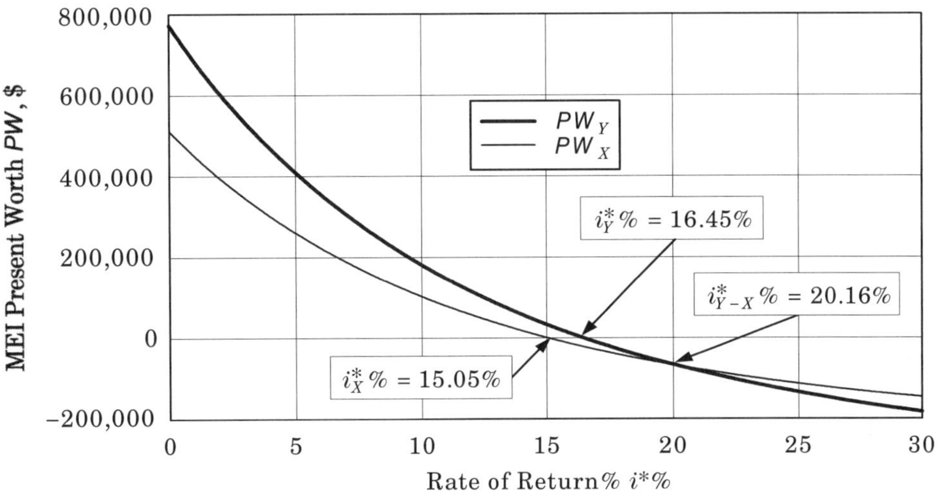

Figure 8.1 The present worth curves of MEIs X and Y in Example 8.1. The ROR% at which they intersect is the IRR% of the incremental investment $Y-X$. NOTE: *The intersection point of the two present worth curves* i^*_{Y-X} *must* ***follow*** *the value of* i^*_Y *when* $i^*_Y > i^*_X$. ***This illustrates Equation (8-3).***

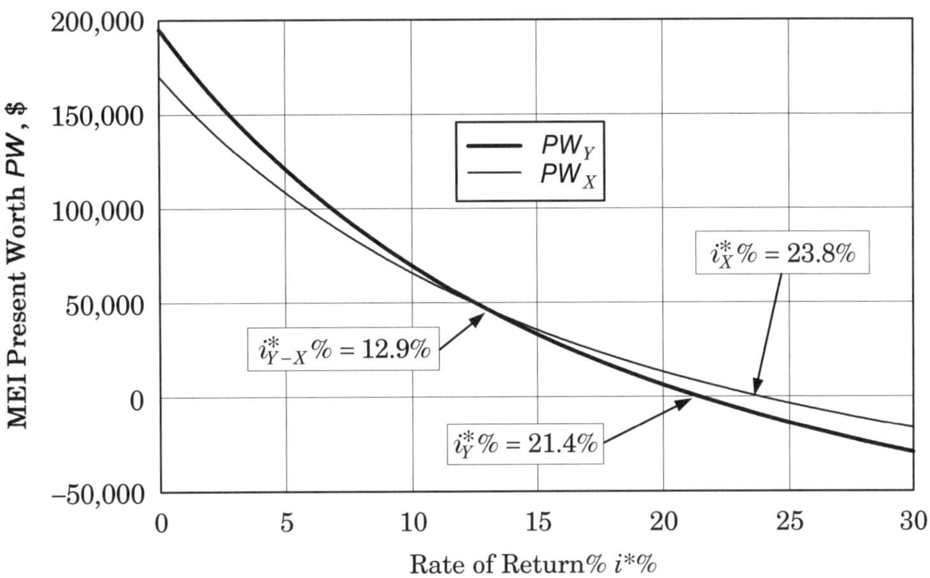

Figure 8.2 The present worth curves of cost reduction MEIs X and Y in Table 8.2. NOTE: *The intersection point of the two present worth curves* i^*_{Y-X} *must* ***precede*** *the value of* i^*_Y *when* $i^*_Y < i^*_X$. ***This illustrates Equation (8-5).***

Computation Note: *Equation (8-4) is true when the present worth curves of MEIs X and Y intersect at a point where their present worths are zero.*

8.2 COMPARING THREE or MORE MUTUALLY EXCLUSIVE INVESTMENTS

8.2.1 General MEI Selection Procedure

To compare n MEIs, where $n \geq 3$, we use Step-by-Step 8.2:

General MEI Selection Procedure: Step-by-Step 8.2
(1) Order the cash flows of the n MEIs by their *increasing* total capitals and calculate their IRRs with Excel's IRR function. If *all* of the IRRs are less than the MARR r^*, *reject all of them*; otherwise go to Step 2.
(2) Make the *first* MEI whose $i^* \geq r^*$ the **defender** $\mathbb{D}$ and the *next* MEI whose $i^* \geq r^*$ the **challenger** $\mathbb{C}$, and Go to Step 3. If there is no *challenger*, **select** the **defender**.
(3) Calculate the IRR of the incremental investment of $\mathbb{C}$ versus $\mathbb{D}$: $i^*_{\mathbb{C}-\mathbb{D}}$.
(4A) If $i^*_{\mathbb{C}-\mathbb{D}} \geq r^*$, make the challenger the defender and determine if a new challenger exists. If it does, Repeat Step 3. *If a new challenger doesn't exist*, **select** the **defender**.
(4B) If $i^*_{\mathbb{C}-\mathbb{D}} < r^*$, determine if a new challenger exists. If it does, go to Step 3. *If a new challenger doesn't exist*, **select** the **defender**.

EXAMPLE 8.2 Finding the Best Pharmaceutical Plant Site (*MEI*-Class)

Problem: A major pharmaceutical company is interested in building a plant in the United States to produce their recently patented cancer drug. Table 8.3 lists *estimates* of the capital disbursements and *aftertax* cash flows of plants located in six different states. The capital disbursements vary because the construction labor wages are different at each site, and the differences in the aftertax cash flows depend primarily on the state income tax rates. *Which plant site, if any, should the company choose when the **aftertax** MARR% is 20% per year?*

TABLE 8.3 Estimated Aftertax Cash Flows of a Pharmaceutical Plant at Sites in Six Different States (*MEI*-Class)[1]

$CF_{Z,k}$, $\$10^6$	−2	−30	−51	24	26	28	30	32	34	34	34	34	44.1
$CF_{Y,k}$, $\$10^6$	−1.6	−27	−47	22	24	26	28	30	32	32	32	32	41
$CF_{X,k}$, $\$10^6$	−1.5	−26	−44	18	20	22	24	26	28	28	28	28	36.5
$CF_{W,k}$, $\$10^6$	−1.3	−25	−43	19	21	23	25	27	29	29	29	29	37.1
$CF_{V,k}$, $\$10^6$	−1.2	−22	−42	18	20	22	24	26	28	28	28	28	35.6
$CF_{U,k}$, $\$10^6$	−1	−20	−40	17	19	21	23	25	27	27	27	27	34
YEAR k	0	1	2	3	4	5	6	7	8	9	10	11	12

[1] Each 12th year cash flow includes the land and working capitals and the salvage value.

Solution: Spreadsheet 8.2 implements Step-by-Step 8.2 for the estimated aftertax cash flows of the six MEIs. It shows that *each MEI will earn more than the MARR% of 20% per year.* Hence, we first make U the defender and V the challenger:

0	A	B	C	D	E	F	G	H
1								
2		***IRR Comparison of MEIs at 6 Plant Locations: r*% = 20%***						
3		***k***	CF_U	CF_V	CF_W	CF_X	CF_Y	CF_Z
4		0	(1)	(1.2)	(1.3)	(1.5)	(1.6)	(2)
5		1	(20)	(22)	(25)	(26)	(27)	(30)
6		2	(40)	(42)	(43)	(44)	(47)	(51)
7		3	17	18	19	18	22	24
8		4	19	20	21	20	24	26
9		5	21	22	23	22	26	28
10		6	23	24	25	24	28	30
11		7	25	26	27	26	30	32
12		8	27	28	29	28	32	34
13		9	27	28	29	28	32	34
14		10	27	28	29	28	32	34
15		11	27	28	29	28	32	34
16		12	34	35.6	37.1	36.5	41	44.1
17		**=IRR**	**29.65%**	**28.94%**	**28.26%**	**26.18%**	**29.19%**	**28.52%**
18		***Calculation of the IRRs of the Incremental Investments***						
19		***k***	CF_{C-D}	CF_{V-U}	CF_{W-U}	CF_{X-U}	CF_{Y-U}	CF_{Z-Y}
20		0		(0.2)	(0.3)	(0.5)	(0.6)	(0.4)
21		1		(2)	(5)	(6)	(7)	(3)
22		2		(2)	(3)	(4)	(7)	(4)
23		3		1	2	1	5	2
24		4		1	2	1	5	2
25		5		1	2	1	5	2
26		6		1	2	1	5	2
27		7		1	2	1	5	2
28		8		1	2	1	5	2
29		9		1	2	1	5	2
30		10		1	2	1	5	2
31		11		1	2	1	5	2
32		12		1.6	3.1	2.5	7	3.1
33			**=IRR**	**17.85%**	**17.70%**	**1.37%**	**27.25%**	**21.17%**
34			***C***	***V***	***W***	***X***	***Y***	***Z***
35			***D***	***U***	***U***	***U***	***U***	***Y***

Spreadsheet 8.2 The top half calculates the IRR%s of the six MEIs and the bottom half calculates the IRR%s of the incremental investments. NOTE: All units are $\$10^6$.

The spreadsheet shows that $i^*_{V-U}\% = 17.85\%$*. Hence, V is unacceptable. The values of* $i^*_{W-U}\%$ and $i^*_{X-U}\%$ *are also less than 20%; hence, W and X are also unacceptable. However, when Y is the challenger:* $i^*_{Y-U}\% = 27.25\%$*. Hence, Y becomes the defender.*

Since $i_Z^*\% = 28.52\%$, we make Z the challenger. The incremental investment $i_{Z-Y}^*\%$ earns 21.17%, and there are no more challengers. Hence, we select MEI Z. It earns 28.52% per year and its incremental investment earns 21.17% per year.

Practice Note: *The logic for choosing Z is that we have the* ***extra*** *7.4 million dollars required to invest in it. If we choose Y, its IRR% will be 0.67% larger than that of Z. But when will another investment come along that will provide an IRR% that is more than 21.17% per year on the* ***extra*** *7.4 million dollars? If we think that these investments come along with the frequency of buses and trains, we should wait.* ***But they don't****, and choosing Z is the correct decision.*

Computation Note: *To help a reader understand your MEI spreadsheet, insert bold borders on the cells containing the IRR and incremental IRR of the* ***selected*** *MEI.*

8.2.2 Relative Values of the Challenger-Defender IRRs

The incremental IRR of a challenger-defender comparison is related to the IRR of the challenger by the Incremental IRR Relations derived in Subsection 8.1.3:

$$\text{If } i_C^* > i_D^*,\ i_{C-D}^* > i_C^*. \tag{8-6}$$

$$\text{If } i_C^* = i_D^*,\ i_{C-D}^* = i_C^*. \tag{8-7}$$

$$\text{If } i_C^* < i_D^*,\ i_{C-D}^* < i_C^*. \tag{8-8}$$

For example, in Spreadsheet 8.2 the first four $C-D$ comparisons are those of the *V, W, X*, and *Y* challengers versus the *U* defender. Since each challenger's value of i_C^* is less than i_U^*, the incremental IRR of each challenger-defender comparison is less than the IRR of the challenger: $i_{C-U}^* < i_C^*$. Similarly, $i_{Z-Y}^* < i_Z^*$, because $i_Z^* < i_Y^*$.

8.2.3 Effect of the MARR on an MEI Study

You saw in Subsection 4.2.6 that the value of the MARR can affect the result of a least-cost study. And so it is with an MEI study. To determine the effect of the MARR on an MEI study, we just compare the MEIs at *different* values of it.

EXAMPLE 8.3 Effect of the MARR on the Selection the Best Pharmaceutical Plant Site (Class 4)

Problem: The manager of the Process Engineering Department of the pharmaceutical company requests that the MEI study be expanded to include aftertax MARR%s of 25% and 30% per year. *Which plant site, if any, should the company choose when the* ***aftertax*** *MARR% is either one of these values. Summarize the results of this example and that in Example 8.2 in a table that will be presented to the chief financial officer (CFO) of the pharmaceutical company.*

Solution: ***MARR% Equals 25% Per Year***

The calculations in Spreadsheet 8.2 remain the same. However, now we select MEI *Y* because the incremental investment $i_{Z-Y}^*\%$ earns only 21.17% per year. MEI *Y* earns 29.19% per year and $i_{Y-U}^*\% = 27.25\%$ per year.

MARR% Equals 30% Per Year

The *incremental* investment calculations in Spreadsheet 8.2 are irrelevant because *none of the six MEIs earns 30% per year.* Hence, *no MEI would be selected.* However, at an MARR% of 29%, MEI U earns 29.65% and MEI Y earns 29.19%. But $i^*_{Y-U}\% = 27.25\%$. Hence, *we would* ***recommend*** *MEI U at an MARR% of 29.65% per year.*

Table 8.4 summarizes the preceding results:

TABLE 8.4 Effect of the MARR% on the Economic Comparison of the Pharmaceutical Plant Locations

MARR% PER YEAR	BEST LOCATION
20%	*Z*
25%	*Y*
29.65%	*U*
30%	None

8.3 EQUAL INITIAL CASH FLOWS

Occasionally, we encounter a comparison of two MEIs when the *initial* negative cash flows are equal. *When this is so, they have no effect on the IRR% of the incremental investment.* Excel ignores the zero value of the initial cash flow of the incremental investment when it calculates the IRR%:

	A	B	C	D	E	F
1						
2		***IRR%s and Incremental IRR% of Two Investments***				
3			***k***	CF_X, $	CF_Y, $	CF_{Y-X}, $
4			0	(15,000)	(15,000)	0
5			1	3,000	2,000	($1,000)
6			2	2,900	3,200	300
7			3	3,000	3,200	200
8			4	3,000	3,200	200
9			5	3,000	3,200	200
10			6	3,000	3,200	200
11			7	3,000	3,200	200
12			8	3,000	3,200	200
13			9	3,000	3,200	200
14			10	3,000	3,200	200
15				**14.97%**	**15.03%**	**16.23%**
16				**=IRR(D4:D14)**	**=IRR(E4:E14)**	**=IRR(F4:F14)**

Spreadsheet 8.3 The zero value in cell F4 has no effect on the IRR%, Excel ignores it.

8.4 SUMMARY

Subsection 8.1.2 explains the basic selection model for two mutually exclusive investments (MEIs). It explains the Incremental IRR Selection Criteria and presents Step-by-Step 8.1 for comparing two MEIs.

Subsection 8.2.1 explains the General MEI Selection Procedure presented in Step-by-Step 8.2 and used for comparing three or more MEIs. Example 8.2 uses Step-by-Step 8.2 to compare six locations for a pharmaceutical plant. Subsection 8.2.3 explains how to determine the effect of the MARR on the conclusion of an MEI study, and Example 8.3 illustrates the procedure.

Section 8.3 shows that when we compare two MEIs with equal initial negative cash flows *they have no effect on the IRR% of the incremental investment.*

8.5 PROBLEMS

(P8-1) Here are the estimated cash flows of six proposed eight-year *mutually exclusive* 1*P-A* cost reduction investments:

TABLE 8.5 Six 1*P-A* Cost Reduction Investments

INVESTMENT	COST, $	ANNUAL SAVINGS, $
A	4,000	800
B	8,000	2,500
C	12,000	4,200
D	16,000	5,000
E	24,000	8,200
F	30,000	9,600

(Part A) *If the MARR% is 15% for this type of investment, which investment, if any, would you recommend?* **(Part B)** *If a MARR% of 20% is acceptable, which investment, if any, would you recommend?*

(P8-2) Table 8.6 shows the estimated cash flows of three proposed mutually exclusive apartment house investments in Phoenix. *Which investment, if any, would you recommend if the aftertax MARR% is 16% per year.*

TABLE 8.6 Estimated Aftertax Cash Flows of Three Mutually Exclusive Apartment House Investments (*MEI*-Class)

$CF_{X,k}$, $\$10^6$	−0.5	−1.0	0.4	0.4	0.4	0.4	0.4	0.4	0.4	0.4	0.4	0.4
$CF_{Y,k}$, $\$10^6$	−0.55	−1.1	0.45	0.45	0.45	0.45	0.45	0.45	0.45	0.45	0.45	0.45
$CF_{Z,k}$, $\$10^6$	−0.90	−1.2	0.6	0.6	0.6	0.6	0.6	0.6	0.6	0.6	0.6	0.6
YEAR *k*	0	1	2	3	4	5	6	7	8	9	10	11

(P8-3) *Add aftertax MARR%s of 18% and 20% per year to the Apartment House Investment comparison, and summarize your results in a table similar to Table 8.4.*

9

Aftertax IRR%

"... *in this world nothing can be said to be certain, except death and **taxes**.*"
Benjamin Franklin, 1789

OBJECTIVES

After reading this chapter you will be able to calculate the aftertax IRR% of an industrial investment without and with a start-up time, ***IT*-Class** problems.

9.1 INCOME TAXES & DEPRECIATION

9.1.1 Combined Income Tax Rate

Companies that earn minimum incomes are required to pay federal and state income taxes. The **income tax** IT_k for year k is the sum of the federal and state income taxes for that year. Since *the state income tax is a deduction in a federal income tax calculation*, the value of IT_k is given by

$$IT_k = r_s TI_{s,k} + r_f (TI_{f,k} - r_s TI_{s,k}), \tag{9-1}$$

where

- r_f and r_s are the **federal** and **state income tax rates**, respectively.
- $TI_{f,k}$ and $TI_{s,k}$ are the **federal** and **state taxable incomes**, respectively.

If we *assume* that the federal and state taxable incomes are equal[1], the right hand side of Equation (9-1) can be simplified:

$$IT_k = \left[r_s + r_f (1 - r_s) \right] TI_k.$$

Hence,

$$\boxed{IT_k = r_c TI_k,} \tag{9-2}$$

where the **combined income tax rate** r_c is defined as

[1] Remember, we are trying to *estimate* the aftertax IRR% of a *proposed* investment whose operating cash flows will normally occur over operating times that range from 5 to 20 years. The uncertainties in the *future* operating cash flows do not warrant a more precise aftertax cash flow model.

$$r_c \equiv r_s + r_f(1 - r_s). \tag{9-3}$$

State tax rates range from 4% to about 8%, and the federal income tax rate on corporate taxable incomes of $18.33 million or more is currently 35%. Hence, without knowing the plant site location, we *assume* a *maximum* combined income tax rate of 40%:

$$r_c = 0.08 + 0.35(1 - 0.08) \approx 0.40. \tag{9-4}$$

Usage Note: *: Obviously, if the federal rate changes or a lower state tax rate exists where the plant will be built, the appropriate combined income tax rate should be used in aftertax IRR% calculations.*

9.1.2 Taxable Income, Income Tax & Aftertax Cash Flow

A company's **taxable income** TI_k for any year k is defined by

$$TI_k \equiv R_k - (PC_k + SC_k + d_k), \tag{9-5}$$

where

- R_k is the receipt for the kth year.
- PC_k is the production cost for the kth year.
- SC_k is the SCORD (Sales Corporate Overhead R & D) cost for the kth year.
- d_k is the depreciation expense for the kth year.

R_k, PC_k, and SC_k are *annual cash flows*, but the **depreciation expense** d_k allocates a part of an investment's fixed capital FC, money *previously* spent, to *reduce* the taxable income for the kth year. ***The value of*** d_k ***is not a cash flow!***

The **income tax** for the kth year is the product of the combined income tax rate r_c and the taxable income:

$$\begin{aligned} IT_k &= r_c TI_k \\ &= r_c\left[R_k - (PC_k + SC_k + d_k)\right]. \end{aligned} \tag{9-6}$$

The **aftertax cash flow** $ATCF_k$ for year k is the difference of the **operating cash flow** OCF_k, which is defined by Equation (6-6), and the income tax IT_k:

$$\begin{aligned} ATCF_k &= \underbrace{R_k - (PC_k + SC_k)}_{OCF_k} - IT_k \\ &= R_k - (PC_k + SC_k + IT_k). \end{aligned} \tag{9-7}$$

9.1.3 Depreciation Expenses

Tax depreciation accounting is a systematic and rational method of allocating an investment's fixed capital as a series of **depreciation expenses** d_k's over the **tax depreciation life** TDL allowed by the Internal Revenue Service (IRS). *The value of TDL is usually smaller than an investment's* ***operating time*** *OT.* For example, the operating time of the chemical plant investment in Subsection 6.2.4 is 10 years, but you will soon see that its tax depreciation life is 6 years.

The **depreciation expense** d_k of an investment in year k of its tax depreciation life is defined by

$$d_k \equiv f_k FC \qquad k = k_{CT+1}, \ldots, k_{CT+TDL}, \tag{9-8}$$

where f_k is the **fractional depreciation** in the kth year of the tax depreciation life and k_{CT} is the value of k at the *end* of the capital time *CT*. The CFD in Figure 9.1 displays the values of *CT* (2a), *TDL* (6a), *OT* (10a) and the *aftertax* cash flows (*ATCF*'s) of the chemical plant investment described in Subsection 6.2.4. A plot plan of the chemical plant is shown in Figure 9.2 on the following page.

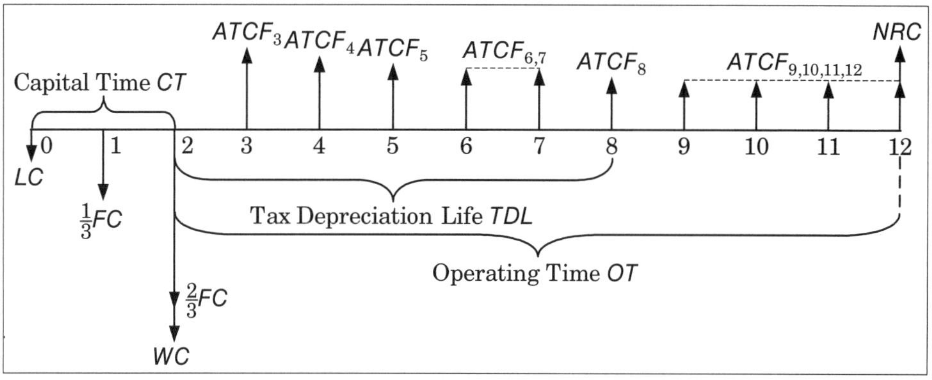

Figure 9.1 ***Aftertax* CFD of the Chemical Plant Investment** NOTE: The cash flow *NRC* represents the **net recovered capital**. It equals the sum of the *land capital (LC)*, *working capital (WC)*, and the *net salvage value* (see Subsection 9.2.2).

The values of *TDL* and the f_k's depend on the asset class of the investment specified by the IRS. It classifies various investments in **Table B-2** of **IRS Publication 946**, a voluminous 114 page document. **Table B-2**[2], which itself spans 9 pages, describes the various IRS asset classes and their **recovery periods** *RP*. The value of any asset's *TDL* is *defined* as one year greater than *RP*:

$$TDL \equiv RP + 1. \tag{9-9}$$

Table 9.1 lists a few of the industrial asset classes and shows the six values of *RP* allowed by the IRS for depreciating *FC*: 3, 5, 7, 10, 15, or 20 years.

TABLE 9.1 Some Common Industrial Asset Classes

ASSET CLASS	DESCRIPTION	*RP*, YEARS
10.0	Mining	7
13.3	Petroleum Refining	10
15.0	Construction	5
28.0	Manufacture of Chemicals	5
32.11	Manufacture of Glass Products-Special Tools	3
32.2	Manufacture of Cement	15
49.13	Electric Utility Steam Production Plant	20

[2] Department of the Treasury, Internal Revenue Service, **Publication 946**, *How To Depreciate Property*, February 27, 2015, pp. 101-109.

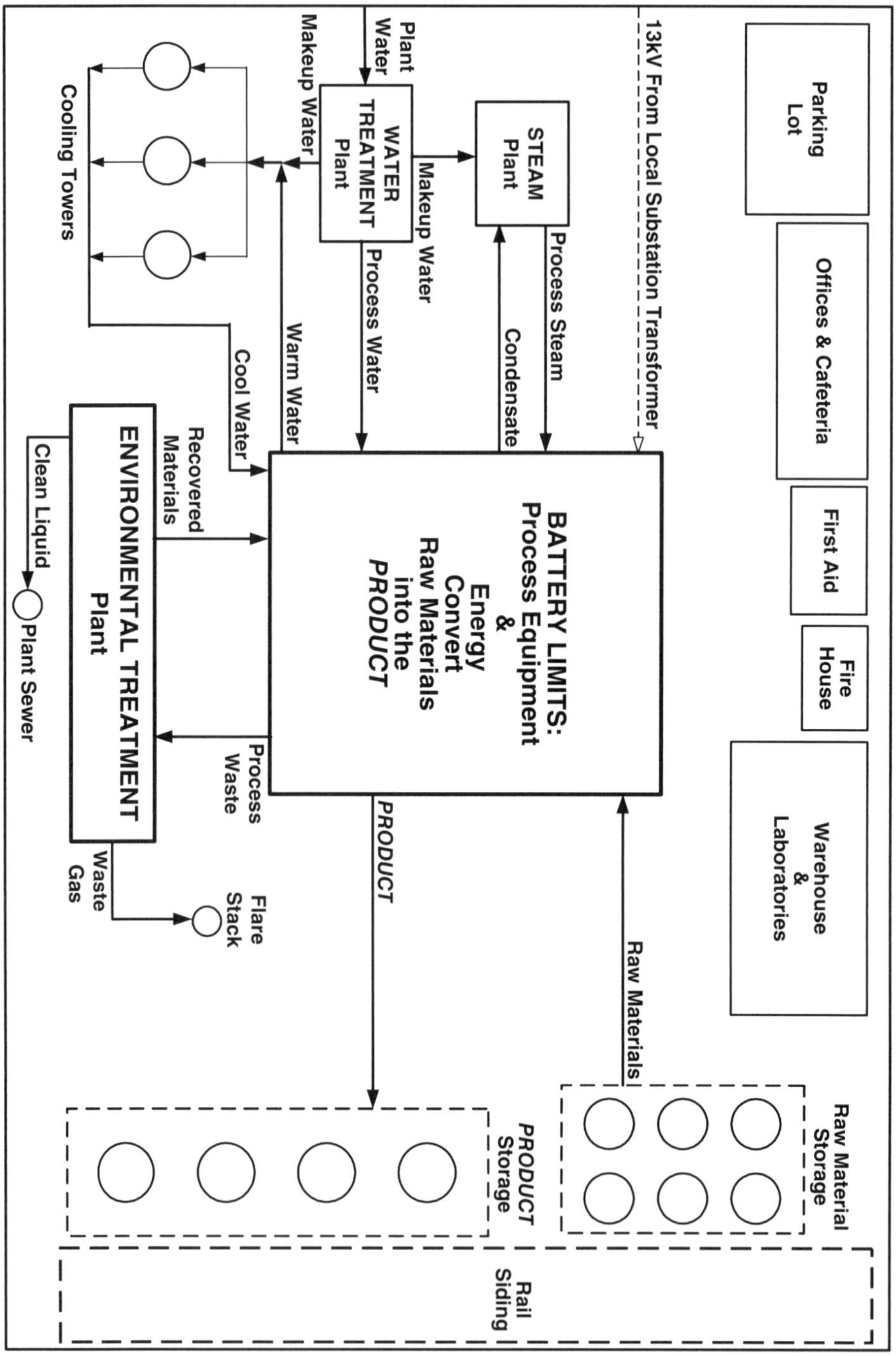

Figure 9.2 **Plot Plan of the Chemical Plant** This outline drawing shows the various components of a chemical plant. From the point-of-view of engineering economics, it's a money factory that converts one or several raw materials into the annual cash flows from the sale of its product(s).

9.2 MACRS DEPRECIATION

9.2.1 Percentage Depreciations

Page 30 in **IRS Publication 946** shows that the **Modified Accelerated Cost Recovery System (MACRS)** is used to depreciate most business assets. To obtain the **percentage depreciations** of the fixed capital with the MACRS, we do the following:

1. Use Table 9.1 or **Table B-2** in **IRS Publication 946** to determine the recovery period of the asset.
2. *Assume* that the asset is placed in service at the *beginning* of its first tax year.
3. Use **Table A-2** in **IRS Publication 946** to determine the **percentage depreciations** ($100f_k$'s) for the asset's recovery period:

Table A-2. **3-, 5-, 7-, 10-, 15-, and 20-Year Property Mid-Quarter Convention Placed in Service in First Quarter**

Year	Depreciation rate for recovery period					
	3-year	5-year	7-year	10-year	15-year	20-year
1	58.33%	35.00%	25.00%	17.50%	8.75%	6.563%
2	27.78	26.00	21.43	16.50	9.13	7.000
3	12.35	15.60	15.31	13.20	8.21	6.482
4	1.54	11.01	10.93	10.56	7.39	5.996
5		11.01	8.75	8.45	6.65	5.546
6		1.38	8.74	6.76	5.99	5.130
7			8.75	6.55	5.90	4.746
8			1.09	6.55	5.91	4.459
9				6.56	5.90	4.459
10				6.55	5.91	4.459
11				0.82	5.90	4.459
12					5.91	4.460
13					5.90	4.459
14					5.91	4.460
15					5.90	4.459
16					0.74	4.460
17						4.459
18						4.460
19						4.459
20						4.460
21						0.565

Department of the Treasury, Internal Internal Revenue Service, **Publication 946**, *How To Depreciate Property*, February 27, 2015, p. 71.

NOTE: *The sum of the percentage depreciations of each asset class is 100%.* Hence, it is *assumed* in the MACRS that ***the salvage value of the investment is zero***.

9.2.2 Net Recovered Capital

If the salvage value of an investment is not zero at the end of the operating time, it is taxable income. Hence, the **net salvage value** NS is recovered at that time:

$$NS = (1 - r_c)S. \tag{9-10}$$

Recovery of the working capital WC is never taxed, and if the land is sold *without* a profit, the land capital LC is recovered. Hence, the **net recovered**

capital *NRC* at the end of the operating time is defined by

$$NRC \equiv LC + WC + (1 - r_c)S. \tag{9-11}$$

9.2.3 Calculating an Aftertax IRR%

Spreadsheet 9.1 on the following page shows the calculation of the *aftertax* IRR% of the chemical plant investment described in Subsection 6.2.4. As you can see, *the annual payments of income taxes have significantly reduced the IRR% from the pretax value of 32.59% to 23.81%.*

The 6 *fractional* depreciations in array H13:H18 are obtained by dividing the 6 percentage depreciations in Table A-2 for the 5-year recovery period of a chemical plant (see Table 9.1) by 100. *The fractional depreciations in subsequent years are all 0.* The annual depreciations are calculated in array I13:I22. The column headings in cells J12 and K12 explain the calculations of the *taxable incomes* and *income taxes* in arrays J13:J22 and K13:K22, respectively. The value of the combined income tax rate r_c is located in cell K11. The aftertax cash flows $ATCF_k$'s are calculated in array M13:M21, and the sum of *NRC* and $ATCF_{12}$ is calculated in cell M22.

Computation Note: *Copies of Spreadsheets 9.1, 9.2, and 9.3 in this chapter are in the Book Templates file. They can be used for investments with different capital times and/or operating times by adding or removing the appropriate rows. They can also be used for other types of assets by using their fractional depreciations.*

9.3 START-UP COST & TIME

9.3.1 Definitions

The start-up of a *new* manufacturing or power plant is not just flipping electrical switches and opening valves and dampers. Despite numerous reviews during the design and construction phases, *some* errors in the plant's design, equipment, materials, and construction remain *undetected* and emerge during the plant start-up. Typical examples are (1) illogical material flow, (2) equipment failure, (3) inadequate equipment design, (4) incorrect equipment and/or instrument installation, and (5) faulty wiring. **Start-up** *S-U* is the work required to correct design and construction errors, and a **start-up time** *S-UT* of 3 months is not uncommon.

An estimate of the **start-up cost** is normally incorporated into the fixed capital *estimate* because of the *inevitability* of some errors in any complex human endeavor. Hence, *our concern is the effect on the IRR% of a 3 month start-up time because it affects the magnitudes of the first and last year aftertax cash flows.*

9.3.2 Percentage Depreciations

Tables A-3, **A-4**, and **A-5** in **IRS Publication 946** provide the percentage depreciations for an asset's recovery period when the asset is placed in service in the second, third, and fourth quarters of the year, respectively. If the start-up time of a manufacturing or power plant is three months, it is placed in service at the *beginning* of the second quarter. **Table A-3** is displayed in problem (P9-1). It shows that for the five year recovery period of the chemical plant investment, the percentage depreciations are 25.00%, 30.00%, 18.00%, 11.37%, 11.37%, and 4.26%.

Spreadsheet 9.1 The payment of income taxes reduces the IRR% from a *pretax* value of 32.59% to an ***aftertax*** value 23.81%.

	A	B	C	D	E	F	G	H	I	J	K	L	M
1													
2				$\$10^6$	***Aftertax IRR%, with MACRS Depreciation,***								
3			FC	69.6	***of the CHEMICAL PLANT INVESTMENT:*** $CT = 2a$ & $OT = 10a$								
4			$WC = 0.15FC$	10.44									
5			$S = 0.05FC$	3.48									
6			PC	147									
7			R	208.8									
8			$SC = 0.15R$	31.32							Note: $ATCF_k = R_k - (PC_k + SC_k + IT_k)$		
9		k										**Capital & ATCFs**	CF
10		0	$LC = 0.01FC$	0.696								LC	(0.696)
11		1	$\frac{1}{3}FC$	23.2						r_c	**0.40**	$\frac{1}{3}FC$	(23.2)
12		2	$\frac{2}{3}FC + WC$	56.84	R	SC	PC	f	$d = fFC$	$TI = R - (PC + SC + d)$	$IT = r_c TI$	$\frac{2}{3}FC + WC$	(56.84)
13		3			208.8	31.32	147	0.35	24.36	6.12	2.45	$ATCF_3$	28.032
14		4			208.8	31.32	147	0.26	18.10	12.38	4.95	$ATCF_4$	25.526
15		5			208.8	31.32	147	0.156	10.86	19.62	7.85	$ATCF_5$	22.631
16		6			208.8	31.32	147	0.11	7.66	22.82	9.13	$ATCF_6$	21.353
17		7			208.8	31.32	147	0.11	7.66	22.82	9.13	$ATCF_7$	21.353
18		8			208.8	31.32	147	0.014	0.96	29.52	11.81	$ATCF_8$	18.672
19		9			208.8	31.32	147	0	0	30.48	12.19	$ATCF_9$	18.288
20		10			208.8	31.32	147	0	0	30.48	12.19	$ATCF_{10}$	18.288
21		11	$NRC = LC + WC + (1 - r_c)S$		208.8	31.32	147	0	0	30.48	12.19	$ATCF_{11}$	18.288
22		12	13.224		208.8	31.32	147	0	0	30.48	12.19	$NRC + ATCF_{12}$	31.512
23				**TOTALS**	**2,088**	**313.2**	**1,470**	**1.0000**	**69.60**	**235.20**	**94.08**	= IRR(M10:M22)	**23.81%**

9.3.3 Calculating the Aftertax IRR% with a Start-Up Time

Spreadsheet 9.2 on the following page shows the calculation of the *aftertax* IRR% of the chemical plant investment when its start-up time is 3 months. As you can see, the IRR% has been reduced from 23.81% to 22.28% because of the less favorable distribution of the aftertax cash flows. However, *the total of the 11 aftertax cash flows is identical to the 10 aftertax cash flows in Spreadsheet 9.1; both equal $\$223.94 \times 10^6$.*

Here is how Spreadsheet 9.2 differs from Spreadsheet 9.1.

1. The values of R, SC, and PC at the end of the 3rd year are 75% of those in Spreadsheet 9.1 because of the 3 month start-up time.
2. The six values of f are different because the plant is placed in service in the *second* quarter of the year. Table A-3 shows that the six percentage depreciations for a recovery period of 5 years are 25.00%, 30.00%, 18.00%, 11.37%, 11.37%, and 4.26% instead of the more favorable ***early*** values in Table A-2: **35%**, **26%**, 15.6%, 11.01%, 11.01% and 1.38%.
3. The operating time is 11 years, and the values of R, SC, and PC at the end of the 13th year are 25% of those in year 10 of Spreadsheet 9.1.

NOTE: All of the column totals are *identical* to those in Spreadsheet 9.1. ***The only difference between the two spreadsheets is the timing of the cash flows, but not their sums.*** This is an another example of ***"the time value of money."***

9.4 STRAIGHT LINE DEPRECIATION

Occasionally **straight line depreciation** is used to calculate the annual depreciations of an investment's fixed capital. With this depreciation method, the **fractional depreciations** are constant throughout the tax depreciation life TDL:

$$f_k \equiv \frac{1}{TDL}. \tag{9-12}$$

When the value of TDL is 6, the 6 fractional depreciations are each 0.1666....

If some or all of the receipts during the depreciation life ***increase*** from year to year, straight line depreciation *may* yield an IRR% that is larger than that obtained with the MACRS. Spreadsheet 9.3 on page 186 shows that the IRR% of the chemical plant investment with straight line depreciation is 23.98% compared to the value of 23.81% with MACRS depreciation in Spreadsheet 9.1. The total values in row 23 of each spreadsheet are *identical*, but the IRR% is larger with straight line depreciation because the aftertax cash flows in years five through eight are much larger than those with MACRS depreciation. This is another example of ***the time value of money***.

Computation Notes: *(1) The percentage MACRS depreciations in Tables A-2, 3, 4, and 5 for 3, 5, 7, and 10 year assets are calculated with the 200% declining balance method with a change to straight line depreciation when it provides an equal or greater deduction.*
(2) The 15 and 20 year assets are calculated with the 150% declining balance method with the same change to straight line depreciation.

Subsection 9.5.2 explains these declining balance depreciations.

Usage Note: *Copies of Spreadsheets 9.1, 9.2, and 9.3 are in the Book Templates file.*

Spreadsheet 9.2 A 3 month start-up time reduces the IRR% from 23.81% to 22.28%.

	A	B	C	D	E	F	G	H	I	J	K	L	M
1													
2				$\$10^6$	*Aftertax IRR%, with MACRS Depreciation and* $S\text{-}UT$ = 0.25a, *of the CHEMICAL PLANT INVESTMENT:* CT = 2a & OT = 11a								
3			FC	69.6									
4			$WC = 0.15FC$	10.44									
5			$S = 0.05FC$	3.48									
6			PC	147									
7			R	208.8									
8			$SC = 0.15R$	31.32							Note: $ATCF_k = R_k - (PC_k + SC_k + IT_k)$		
9		k										Capital & ATCFs	CF
10		0	$LC = 0.01FC$	0.696								LC	(0.696)
11		1	$\frac{1}{3}FC$	23.2						r_c	0.40	$\frac{1}{3}FC$	(23.2)
12		2	$\frac{2}{3}FC+WC$	56.84	R	SC	PC	f	$d = fFC$	$TI = R - (PC + SC + d)$	$IT = r_c TI$	$\frac{2}{3}FC+WC$	(56.84)
13		3			156.6	23.49	110	0.25	17.40	5.46	2.18	$ATCF_3$	20.676
14		4			208.8	31.32	147	0.3	20.88	9.60	3.84	$ATCF_4$	26.640
15		5			208.8	31.32	147	0.18	12.53	17.95	7.18	$ATCF_5$	23.299
16		6			208.8	31.32	147	0.114	7.91	22.57	9.03	$ATCF_6$	21.453
17		7			208.8	31.32	147	0.114	7.91	22.57	9.03	$ATCF_7$	21.453
18		8			208.8	31.32	147	0.043	2.96	27.52	11.01	$ATCF_8$	19.474
19		9			208.8	31.32	147	0	0	30.48	12.19	$ATCF_9$	18.288
20		10			208.8	31.32	147	0	0	30.48	12.19	$ATCF_{10}$	18.288
21		11			208.8	31.32	147	0	0	30.48	12.19	$ATCF_{11}$	18.288
22		12	$NRC = LC + WC + (1 - r_c)S$		208.8	31.32	147	0	0	30.48	12.19	$ATCF_{12}$	18.288
23		13	13.224		52.2	7.83	37	0	0	7.62	3.05	$NRC + ATCF_{13}$	17.796
24		14		**TOTALS**	**2,088**	**313.2**	**1,470**	**1.0000**	**69.60**	**235.20**	**94.08**	= IRR(M10:M23)	**22.28%**

Spreadsheet 9.3 Using straight line depreciation instead of the MACRS ***increases*** the IRR% from 23.81% to 23.98%.

	A	B	C	D	E	F	G	H	I	J	K	L	M
1													
2				$\$10^6$	***Aftertax IRR%, with Straight Line Depreciation, of the CHEMICAL PLANT INVESTMENT: CT = 2a & OT = 10a***								
3			**FC**	69.6									
4			$WC = 0.15FC$	10.44									
5			$S = 0.05FC$	3.48									
6			**PC**	147									
7			R	208.8									
8			$SC = 0.15R$	31.32							Note: $ATCF_k = R_k - (PC_k + SC_k + IT_k)$		
9		k										**Capital & ATCFs**	**CF**
10		0	$LC = 0.01FC$	0.696								**LC**	(0.696)
11		1	$\frac{1}{3}FC$	23.2						r_c	**0.40**	$\frac{1}{3}FC$	(23.2)
12		2	$\frac{2}{3}FC+WC$	56.84	R	**SC**	**PC**	f	$d = fFC$	$TI = R - (PC + SC + d)$	$IT = r_c TI$	$\frac{2}{3}FC+WC$	(56.84)
13		3			208.8	31.32	147	0.167	11.60	18.88	7.55	$ATCF_3$	22.928
14		4			210.8	31.32	147	0.167	11.60	20.88	8.35	$ATCF_4$	24.128
15		5			212.8	31.32	147	0.167	11.60	22.88	9.15	$ATCF_5$	25.328
16		6			214.8	31.32	147	0.167	11.60	24.88	9.95	$ATCF_6$	26.528
17		7			216.8	31.32	147	0.167	11.60	26.88	10.75	$ATCF_7$	27.728
18		8			214.8	31.32	147	0.167	11.60	24.88	9.95	$ATCF_8$	26.528
19		9			204.8	31.32	147	0	0	26.48	10.59	$ATCF_9$	15.888
20		10			201.8	31.32	147	0	0	23.48	9.39	$ATCF_{10}$	14.088
21		11	$NRC = LC + WC + (1 - r_c)S$		201.8	31.32	147	0	0	23.48	9.39	$ATCF_{11}$	14.088
22		12	13.224		200.8	31.32	147	0	0	22.48	8.99	$NRC + ATCF_{12}$	26.712
23				**TOTALS**	**2,088**	**313.2**	**1,470**	**1.0000**	**69.60**	**235.20**	**94.08**	**= IRR(M10:M22)**	**23.98%**

9.5 BOOK DEPRECIATION METHODS

In addition to income tax depreciation, a company will depreciate an asset over its *life* L[3] in order to recognize the decrease in its monetary value due to wear and tear. This **annual depreciation** is called **book depreciation** d_B, and the **book value** (BV_B) of an asset at the *end* of the kth year of its life is *defined* as the difference of its **cost** C and the sum of its book depreciations:

$$BV_B(k) \equiv C - \sum_{y=1}^{k} d_B(y), \tag{9-13}$$

where y represents *every* year up to and including the kth year. This section explains two book depreciation methods that are the basis of *most* MACRS depreciations: **straight line** and **double declining balance**. It also explains **sum-of-years-digits** depreciation; it was often used for income tax depreciation from 1954 to 1981.

9.5.1 Straight Line Depreciation

*If an asset's book value decreases linearly over its life, the constant annual (book) depreciation is called **straight line (SL) depreciation:***

$$d_{SL}(y) \equiv (C - S)L^{-1}. \tag{9-14}$$

In Equation (9-14)

- C is the cost of the asset, \$.
- S is its salvage value at the end of its life, \$.
- L is the ***numeric*** of the life measured in years, dimensionless.

Since the annual depreciations are constant, Equation (9-13) reduces to

$$BV_{SL}(k) = C - \sum_{y=1}^{k} d_{SL}(y) = C - k(C - S)L^{-1}. \tag{9-15}$$

EXAMPLE 9.1 Calculating the SL Book Values of the Chemical Plant

Problem: *Calculate the 10 straight line book values of the cost C—fixed capital FC—of the chemical plant investment in two ways: (1)* ***Pocket Calculator (PC)*** *and (2)* ***Excel***. NOTE: The land and working capitals do not depreciate.

(PC) Solution: In Spreadsheet 9.1 the fixed capital is $\$69.6\times10^6$, the salvage value S is $\$3.48\times10^6$, and L is the *numeric* (10) of the 10 year operating time. The book value at the end of each year k in the life is given by Equation (9-15):

$$\begin{aligned} BV_{SL}(k) &= C - k(C - S)L^{-1} \\ &= \left[69.6 - k(69.6 - 3.48)10^{-1}\right]\times\$10^6 \\ &= (69.6 - 6.612k)\times\$10^6 \quad k = 1, 2, \dots, 10. \end{aligned} \tag{E9.1-1}$$

---

[3]An assest's **life** is the timespan on which it meets its performance specifications. See Subsection 4.1.1.

The 10 book values are calculated in Table 9.2:

TABLE 9.2 SL Book Values of the Chemical Plant

k	$BV_{SL}(k) = (69.6 - 6.612k) \times \$10^6$
1	62.988
2	56.376
3	49.764
4	43.152
5	36.540
6	29.928
7	23.316
8	16.704
9	10.092
10	$S = 3.48$

Excel Solution: Spreadsheet 9.4 displays the calculations. The Excel function **SLN(Cost, Salvage Value, Life)** calculates the annual straight line depreciation over the investment's life. The formula in merged cell D2—SLN(F2,H2,J2)—calculates the *value* of the annual depreciation: $6.612 \times \$10^6$. The 10 book values in array C4:L4 are obtained by pasting the book value \$F2-\$D2 * C3 in cell C4 across array D4:L4. They are identical to those in Table 9.2.

	A	B	C	D	E	F	G	H	I	J	K	L
1				d_{SL}, \$$10^6$		C, \$$10^6$		S, \$$10^6$		L		
2				6.612		69.6		3.48		10		
3			k	1	2	3	4	5	6	7	8	9	10
4		$BV_{SL}(k)$, \$$10^6$	62.99	56.38	49.76	43.15	36.54	29.93	23.32	16.70	10.09	3.48

Spreadsheet 9.4 ***Straight Line Book Values of the CHEMICAL PLANT***

The straight line *fractional* depreciations for *United States income taxes* are based on two *assumptions*:

1. The salvage value is zero.
2. The life L is replaced by the IRS tax depreciation life TDL for the asset.

Hence,

$$f_{SL}(y) \equiv \frac{d_{SL}(y)}{C} \equiv \frac{1}{TDL}. \tag{9-16}$$

9.5.2 Double Declining Balance Depreciation

The values of many assets, such as trucks, construction equipment, power and chemical plants, and compressors, do not decrease linearly with age because they become more costly to operate and maintain. Hence, *straight line depreciation is **not** a realistic measure of the book values of these assets.* To correct this, **double declining balance** (DDB) depreciation is often used to depreciate them:

$$f_{\text{DDB}} \equiv \frac{d_{\text{DDB}}(y)}{C} \equiv \frac{2}{L}. \tag{9-17}$$

This is *twice the straight line depreciation rate when the salvage value is zero.* The yearly book values are *defined* by the following equations:

$$\begin{aligned} BV_{\text{DDB}}(1) &\equiv C(1 - f_{\text{DDB}}) \\ BV_{\text{DDB}}(2) &\equiv BV_{\text{DDB}}(1)(1 - f_{\text{DDB}}) \\ \vdots \qquad & \qquad \vdots \\ BV_{\text{DDB}}(k) &\equiv BV_{\text{DDB}}(k-1)(1 - f_{\text{DDB}}). \end{aligned} \tag{9-18a,b,k}$$

Computation Notes: *(1) The book value at the* ***end*** *of any year is* $(1 - f_{\text{DDB}})$ *times the book value at the* ***end*** *of the* ***preceding*** *year.*
(2) A DDB book value rarely equals the salvage value.
(3) ***To obtain a book value equal to the salvage value, we switch to straight line depreciation in the first year that the DDB depreciation equals or is less than that with straight line depreciation.***

EXAMPLE 9.2 Calculating the DDB & SL Book Values of the Chemical Plant with a Pocket Calculator (PC)

Problem: *Use a* ***(PC)*** *to obtain the 10 DDB & SL book values of the cost of the chemical plant.* NOTE: Using a **(PC)** clearly illustrates the DDB & SL switching strategy. Example 9.3 uses Excel for the calculations.
Solution: We first calculate the DDB fractional depreciation with Equation (9-17):

$$f_{\text{DDB}} = \frac{2}{L} = \frac{2}{10} = 0.2.$$

The book value at the *end* of the first year is given by Equation (9-18a):

$$\begin{aligned} BV_{\text{DDB}}(1) &= C(1 - f_{\text{DDB}}) \\ &= \$69.6 \times 10^6 (1 - 0.2) = \$55.68 \times 10^6. \end{aligned}$$

Hence, the depreciation for the year is given by

$$d_{\text{DDB}}(1) = BV_{\text{DDB}}(0) - BV_{\text{DDB}}(1) = (69.6 - 58.68) \times \$10^6 = \$13.92 \times 10^6.$$

To organize the remaining DDB calculations, we put them in a table:

TABLE 9.3 DDB Depreciation of the Chemical Plant Investment

Year k	BV_{k-1}, \$$10^6$	$BV_k = BV_{k-1}(1 - f_{\text{DDB}})$, \$$10^6$	$d_{\text{DDB},k}$, \$$10^6$	$d_{\text{SL},k}$, \$$10^6$
1	69.6	55.68	13.92	6.612
2	55.68	44.544	11.136	6.612
3	44.544	35.635	8.909	6.612
4	35.635	28.508	7.127	6.612
5	28.508	22.807	**5.702**	6.612

In the 5th year the DDB depreciation is $\$5.702\times\$10^6$, which is less than the straight line value of $\$6.612\times\$10^6$. Hence, we ***switch*** to straight line depreciation in the 5th year with a salvage value of $\$3.48\times\$10^6$. Here is the *annual straight line* depreciation that will yield a *straight line* book value of $BV_{SL}(10)$ that equals $\$3.48\times\$10^6$:

$$d_{SL} = \frac{BV_{SL}(4) - BV_{SL}(10)}{6} \equiv \frac{(28.508 - 3.48)\times\$10^6}{6} = \$4.1713\times10^6.$$

Hence,

$$\begin{aligned}
BV_{SL}(5) &= BV_{SL}(4) - d_{SL} = (28.508 - 4.1713)\times\$10^6 = \$24.337\times10^6\\
BV_{SL}(6) &= BV_{SL}(5) - d_{SL} = (24.337 - 4.1713)\times\$10^6 = \$20.165\times10^6\\
BV_{SL}(7) &= BV_{SL}(6) - d_{SL} = (20.165 - 4.1713)\times\$10^6 = \$15.994\times10^6\\
BV_{SL}(8) &= BV_{SL}(7) - d_{SL} = (15.994 - 4.1713)\times\$10^6 = \$11.823\times10^6\\
BV_{SL}(9) &= BV_{SL}(8) - d_{SL} = (11.823 - 4.1713)\times\$10^6 = \$7.651\times10^6\\
BV_{SL}(10) &= BV_{SL}(9) - d_{SL} = (7.651 - 4.1713)\times\$10^6 = \underline{\underline{\$3.48\times10^6}}.
\end{aligned}$$

EXAMPLE 9.3 Calculating the DDB & SL Book Values of the Chemical Plant with Excel

Problem: *Use Excel's DDB and SLN functions to obtain the 10 DDB & SL book values of the cost of the chemical plant investment.* NOTE: The **DDB(Cost, Salvage Value, Life, *k*)** function calculates the DDB depreciation in year *k* of the DDB part of the depreciation life, and the SLN function calculates the straight line depreciation in the SL part of the depreciation life.

Solution: Spreadsheet 9.5 on the following page displays the Excel calculations. As expected, the results are identical to those Example 9.2.

9.5.3 Sum-of-Years-Digits Depreciation

This rapid depreciation method is based on an interesting property of *any* sum of the *first N* positive integers: the **sum-of-years-digits** (*SOYD*) is given by

$$\begin{aligned}
SOYD &\equiv \sum_{k=1}^{N} k = 1 + 2 + \cdots + (N-1) + N\\
&= \frac{N(N+1)}{2}.
\end{aligned} \tag{9-19}$$

For example, if $N = 6$, $SOYD = 1 + 2 + 3 + 4 + 5 + 6 = 21$. Using Equation (9-19) yields the same result:

$$SOYD = \frac{N(N+1)}{2} = \frac{6(6+1)}{2} = 21.$$

If an investment has a cost *C*, salvage value *S*, and life *L*, SOYD depreciations are *defined* by Equations (9-20a,b,k):

	A	B	C	D	E	F
1						
2			***C*, \$10^6**	69.6	Cell D8 contains =DDB(D\$2,D\$3,\$D4,B8) Copy it and Paste Down Array D9:D12	
3			***S*, \$10^6**	3.48		
4			***L***	10		
5						
6		***k***	**$BV_{DDB}(k)$, \$10^6**	**DDB Depreciation**		
7		0	69.6	**$d_{DDB}(k)$,\$10^6**	**$d_{SL}(k)$,\$10^6**	
8		1	55.680	13.920	6.612	
9		2	44.544	11.136	6.612	
10		3	35.635	8.909	6.612	
11		4	28.508	7.127	6.612	
12		5	22.807	**5.702**	6.612	
13				**SL Depreciation**	**$d_{SL}(k)$,\$10^6**	
14			**$BV_{SL}(k)$, \$10^6**	=SLN(C11,D3,6)	4.1714	
15		5	24.337	=C11-E\$14		
16		6	20.165	=C15-E\$14		
17		7	15.994	Copy Cell C16 and Paste it Down Array C17:C20		
18		8	11.823			
19		9	7.651			
20		10	3.48			

Spreadsheet 9.5 ***DDB & SL Book Values of the CHEMICAL PLANT***

***Computation Note:**: A 150% declining balance depreciation rate is used by the IRS to provide the depreciation tables for 15 and 20 year assets.*

$$
\begin{aligned}
d_{\mathrm{SOYD}}(1) &\equiv L\Delta \\
d_{\mathrm{SOYD}}(2) &\equiv (L-1)\Delta \\
\vdots \quad & \quad \vdots \\
d_{\mathrm{SOYD}}(k) &\equiv [L-(k-1)]\Delta,
\end{aligned}
\tag{9-20a,b,k}
$$

where

$$\Delta \equiv \frac{C-S}{SOYD}. \tag{9-21}$$

As required, the sum of the L depreciations equals $C-S$:

$$\sum_{k=1}^{L} d_{\mathrm{SOYD}}(k) = \Delta \sum_{k=1}^{L} k = \Delta \times SOYD = C-S. \tag{9-22}$$

The book value at the end of the kth year is obtained with Equation (9-23):

$$BV_{\text{SOYD}}(k) = C - \sum_{y=1}^{k} d_{\text{SOYD}}(y). \tag{9-23}$$

EXAMPLE 9.4 Calculating the SOYD Book Values of the Chemical Plant

Problem: *Calculate the 10 SOYD book values of the cost C—fixed capital FC—of the chemical plant investment in two ways: (1)* ***Pocket Calculator (PC)*** *and (2)* ***Excel****.*

(PC) Solution: We first calculate the values of *SOYD* and Δ because all SOYD depreciations depend on them:

$$SOYD = \frac{L(L+1)}{2} = \frac{10 \times 11}{2} = 55$$

$$\Delta \equiv \frac{C-S}{SOYD} = \frac{(69.6-3.48)\times \$10^6}{55} = 1.202 \times \$10^6.$$

The following 10 sequential calculations return the 10 SOYD book values of *C*.

$$BV_{\text{SOYD}}(1) = C - L\Delta = (69.6 - 10\times 1.202)\times \$10^6 = \$57.58\times 10^6$$
$$BV_{\text{SOYD}}(2) = BV_{\text{SOYD}}(1) - (L-1)\Delta = (57.58 - 9\times 1.202)\times \$10^6 = \$46.76\times 10^6$$
$$BV_{\text{SOYD}}(3) = BV_{\text{SOYD}}(2) - (L-2)\Delta = (46.76 - 8\times 1.202)\times \$10^6 = \$37.14\times 10^6$$
$$BV_{\text{SOYD}}(4) = BV_{\text{SOYD}}(3) - (L-3)\Delta = (37.14 - 7\times 1.202)\times \$10^6 = \$28.73\times 10^6$$
$$BV_{\text{SOYD}}(5) = BV_{\text{SOYD}}(4) - (L-4)\Delta = (28.73 - 6\times 1.202)\times \$10^6 = \$21.52\times 10^6$$
$$BV_{\text{SOYD}}(6) = BV_{\text{SOYD}}(5) - (L-5)\Delta = (21.52 - 5\times 1.202)\times \$10^6 = \$15.50\times 10^6$$
$$BV_{\text{SOYD}}(7) = BV_{\text{SOYD}}(6) - (L-6)\Delta = (15.50 - 4\times 1.202)\times \$10^6 = \$10.69\times 10^6$$
$$BV_{\text{SOYD}}(8) = BV_{\text{SOYD}}(7) - (L-8)\Delta = (10.69 - 3\times 1.202)\times \$10^6 = \$7.09\times 10^6$$
$$BV_{\text{SOYD}}(9) = BV_{\text{SOYD}}(8) - (L-9)\Delta = (7.09 - 2\times 1.202)\times \$10^6 = \$4.68\times 10^6$$
$$BV_{\text{SOYD}}(10) = BV_{\text{SOYD}}(9) - (L-10)\Delta = (4.68 - 1.202)\times \$10^6 = \$3.48\times 10^6.$$

Excel Solution: Spreadsheet 9.6 eliminates the tedious **(PC)** calculations.

	A	B	C	D	E	F	G	H	I	J	K	L	M
1				***C*, \$10⁶**		***S*, \$10⁶**		***L***	***SOYD***	**Δ, \$10⁶**			
2				69.6		3.48		10	55	1.202			
3			***k***	1	2	3	4	5	6	7	8	9	10
4			**$d_{\text{SOYD}}(k)$**	12.02	10.82	9.62	8.42	7.21	6.01	4.81	3.61	2.40	1.20
5		**$BV_{\text{SOYD}}(k)$, \$10⁶**		57.58	46.76	37.14	28.73	21.51	15.50	10.69	7.09	4.68	3.48

Spreadsheet 9.6 ***SOYD Book Values of the CHEMICAL PLANT***

Cell D4 calculates $d_{\text{SOYD}}(1)$ with the formula =H2*J2, which represents Equation (9-20a): $L\Delta$. The formula =($H2-D3)*$J2 in E4 calculates $d_{\text{SOYD}}(2)$. Pasting this formula across array F4:M4 calculates the remaining 8 depreciation values $d_{\text{SOYD}}(3)....d_{\text{SOYD}}(10)$.

The formula =D2-D4 in cell D5 calculates $BV_{\text{SOYD}}(1)$, and the formula =D5-E4 in cell E5 calculates $BV_{\text{SOYD}}(2)$. Pasting this formula across array F5:M5 calculates the remaining 8 book values.

9.6 BRIEF HISTORY OF U.S. TAX LAWS

Allowable income tax depreciation methods and lives have varied over the years. The income tax laws of 1934, 1954, 1962, 1971, 1981, and 1986 reflect attempts by the U.S. government to strike a balance between fairness, simplicity, and common sense. For example, prior to 1934 companies would often use depreciation lives that were *much* smaller than their corresponding service lives.[4] If the pretax cash flows were essentially uniform, this resulted in *inordinately* small early aftertax cash flows and, hence, *inordinately* small early income taxes. (*These companies clearly understood the time value of money!*) To correct this, the 1934 tax law required that the depreciation life should equal the service life.

This requirement was gradually relaxed over the years. For example, IRS Publication 534 for 1981 states the following: ***"Before 1981 you could use any reasonable method (of depreciation) for every kind of depreciable property."*** The *rapid* **Sum-of-Years Digits** (SOYD) depreciation method was authorized in the 1954 tax law[5] and was used until the 1981 tax law was enacted. It required the use of the **Accelerated Cost Recovery System** (**ACRS**), which was replaced by the **Modified Accelerated Cost Recovery System (MACRS)** in the Tax Reform Act of 1986.

9.7 SUMMARY

Subsection 9.1.1 defines the **combined federal and state income tax rate**, and Subsection 9.1.2 defines the **taxable income**, **income tax**, and **aftertax cash flow** for any year. Subsection 9.1.3 explains the elements of depreciation accounting: **depreciation expense** d_k, **depreciation life** DL, **fractional depreciation** f_k, and **recovery period** RP. Table 9.1 lists some common industrial asset classes and their recovery periods.

Subsection 9.2.1 explains the use of the **MACRS** for depreciating the fixed capital of an industrial asset, and **IRS Table A-2** specifies the percentage depreciations for assets placed in service in the first quarter of the year according to their recovery periods. Subsection 9.2.2 explains **net recovered capital** NRC, and Subsection 9.2.3 explains the MACRS calculation of the IRR% in Spreadsheet 9.1.

Section 9.3 explains **start-up cost** and **start-up time**, and Spreadsheet 9.2 illustrates how a start-up time reduces the IRR%. **Straight line depreciation** is explained in Section 9.4, and its use is illustrated in Spreadsheet 9.3.

Subsections 9.5.1 through 9.5.3 explain straight line, double declining balance, and SOYD **book depreciations**, respectively. Examples 9.1 through 9.4 illustrate their use in calculating the book depreciations of the chemical plant.

9.8 PROBLEMS

(P9-1) A professional engineer hired by a glass products company reported that there will be a shortage in the manufacturing capacity required to satisfy the increased demand for the *special* tools used in the production of glass products.

[4]Grant, E.L., Ireson, W.G., and Leavenworth, R.S. *Principles of Engineering Economy*, 7th ed., p. 200. New York: John Wiley and Sons, 1982.

[5]Op. Cit., p.204.

The consultant also provided Table 9.4. It contains the economic data for a new plant that would satisfy the increased demand.

TABLE 9.4 Economic Data for a Plant to Manufacture Glass Production Tools

ECONOMIC DATA	
Land Capital, *LC*	$30,000
Fixed Capital, *FC*	$4,750,000
Salvage Value, *S*	$200,000
Working Capital, *WC*	$350,000
Production Costs, *PC*	$7,000,000
SCORD Costs, *SC*	$1,470,000
Receipts, *R*	$9,800,000
Construction Time, *CT*	1a
Operating Time, *OT*	8a

The start-up date is the *beginning* of the 2nd quarter of a tax year, and the anticipated operating time is 8 years. **Table A-3** (shown below) in **IRS Publication 946** lists the percentage depreciations for this start time. *What is the aftertax IRR% of this investment if the combined income tax rate is 35%?*

Table A-3. **3-, 5-, 7-, 10-, 15-, and 20-Year Property Mid-Quarter Convention Placed in Service in Second Quarter**

Year	Depreciation rate for recovery period					
	3-year	5-year	7-year	10-year	15-year	20-year
1	41.67%	25.00%	17.85%	12.50%	6.25%	4.688%
2	38.89	30.00	23.47	17.50	9.38	7.148
3	14.14	18.00	16.76	14.00	8.44	6.612
4	5.30	11.37	11.97	11.20	7.59	6.116
5		11.37	8.87	8.96	6.83	5.658
6		4.26	8.87	7.17	6.15	5.233
7			8.87	6.55	5.91	4.841
8			3.34	6.55	5.90	4.478
9				6.56	5.91	4.463
10				6.55	5.90	4.463
11				2.46	5.91	4.463
12					5.90	4.463
13					5.91	4.463
14					5.90	4.463
15					5.91	4.462
16					2.21	4.463
17						4.462
18						4.463
19						4.462
20						4.463
21						1.673

Department of the Treasury, Internal Internal Revenue Service, **Publication 946**, *How To Depreciate Property*, February 27, 2015, p. 72.

(P9-2) Calculate with Excel the straight line and double declining balance depreciations of a \$230,000 electric generator (with a \$20,000 salvage value) on its 8 year life.

(P9-3) Calculate the SOYD depreciation with Excel for the following investment:

***CF*, \$10³**	–1	–1	2	2	2	3	3
YEAR *k*	0	1	2	3	4	5	6

10

Sensitivity Analysis

"U.S. chemical companies are involved in too many enterprises considerably less profitable than seemed likely at the time investment decisions were made. Two factors, ***sales price*** *and* ***unit sales volume****, are likely to be the* ***principle troublemakers****. No consideration is usually given to other competing companies' technical and marketing potential to produce and sell the same chemical or competing chemicals to the user."*
Lloyd Johnson, 1962

"The return on investment is far more sensitive to selling price, but an enormous effort is spent refining the investment estimate, while the ***final sales price*** *and* ***sales volume*** *figures more resemble* ***guesstimates*** *than the best possible estimates."*
Richard Strauss, 1968

"Some of the major oil and chemical firms after their recent capacity expansion found a few years later that ***demand decreased*** *and their competitors also* ***increased capacity****. Obviously, many of these firms* ***failed to properly consider at the identification phase the actions of competitors and the economy.*** *Otherwise, some of these projects could have been quickly disposed of without spending a lot of time, effort, and money only to:* ***(1) cancel them later on; or (2) achieve less than the desired results when the projects were completed.****"*
George E. Pinches, 1984

"Those who cannot remember the past are condemned to repeat it."
George Santayana, 1905

OBJECTIVES

It should be clear from the above quotations that some companies may fail to consider the effects of changes in the *expected* sales volume and product price on the IRR% of a *proposed* investment. If the investment is funded, *unexpected* reductions in the annual receipts will lead to a smaller IRR% and, possibly, termination of the investment.

In any event, we must incorporate a **sensitivity analysis** into our IRR% **calculations** to obtain a *realistic* estimate of the effect of the annual receipts (or another input variable) on the IRR% of a *proposed* investment. After reading this *short* chapter you will be able do this.

You will also be able perform a **sensitivity analysis** on **staged construction**.

10.1 CREATING an IRR% SENSITIVITY TABLE

We can use any existing spreadsheet that calculates an IRR%, such as Spreadsheet 9.1 in the *Book Templates* file, to calculate the effect on the IRR% of changing the value of *any* input variable, such as *FC*, *PC*, or *R*. If we do this for several values of *R*, we can manually create a **Sensitivity Table** of the values of the *IRR%* versus the value of *R*.

However, we can conveniently create the Sensitivity Table *in* the spreadsheet by creating the corresponding **Data Table** in it. We use Step-by-Step 10.1 to do this:

Creating a Data Table in a Spreadsheet: Step-by-Step 10.1

(1) Create an ***empty*** **Data Table** in the spreadsheet. Spreadsheet 10.1 on the following page is a copy of Spreadsheet 9.1 with a different title and an *unevaluated* **Data Table** in array G4:H13.

(2A) Type the *name* of the variable that we want to vary, *R*, in cell G4.

(2B) Copy the *formula* that depends on *R* and calculates the *IRR%*; it is in cell M25. Note: Use *absolute* addresses:

=IRR($M$12:$M$24).

(2C) Select the empty array H4:H13 and **paste** this *formula* into it. The evaluated formula value 23.81% is displayed in each cell.

(3) Array G5:G13 lists the values we want to assign to the annual receipt *R*. We *calculate* them from the specified **percentage changes** in **array F5:F13** of the *default* value of *R*. For example, the value in cell G5, 167.0, is obtained from

$$= \underbrace{(100\% + \text{F5})}_{100\%-20\%} * \underbrace{\$\text{D}\$7}_{208.8} = 167.04 = 167.0.$$

Paste this formula into array G6:G13 to complete the array of *R* values.

Computation Note: *Excel interprets the product of a percentage times a number as (percentage / 100) times the number. The numbers in array G5:G13 in Spreadsheet are* <u>*displayed*</u> *with one decimal place.*

(4a) To evaluate the **Data Table**, select array (G4:H13) and *Click* the **Data** tab. In the **Data Tools** group *Click* **What-If Analysis** >*Click* **Data Table...** to open the **Data Table...** dialog box. In the **Row input cell:** box[1] *Click* the cell containing the value of *R,* which is D7, > *Click* **OK**.

(4b) Excel evaluates the value of the IRR% for each value of *R* in the **Data Table.** Spreadsheet 10.2 displays the evaluated **Data Table**. NOTE: When $\Delta R\% = 0$, the value of the *IRR%* is 23.81%, which is the calculated value in cell M25.

(5) Figure 10.1, which follows Spreadsheet 10.2, shows the variation of the IRR% with the percentage change in the value of *R* from its default value of $\$208.8 \times 10^6$. ***This information is mandatory for the investment decision maker.***

[1] We must use the **Row Input Cell:** box because Excel will *calculate* the values of the rows of *R* *and* the IRR% when it evaluates the Data Table.

Spreadsheet 10.1 ***Creation of an unevaluated IRR% versus R Data Table for the Chemical Plant Investment***

	A	B	C	D	E	F	G	H	I	J	K	L	M
1													
2				$\$10^6$	(IRR% *versus R*)$_{\text{unevaluated}}$ of the CHEMICAL PLANT INVESTMENT: *CT*= 2a, MACRS								
3			FC	69.6			Data Table						
4			$WC = 0.15FC$	10.44		$\Delta R\%$	R	23.81%					
5			$S = 0.05FC$	3.48		-20%	167.0	23.81%					
6			PC	147		-15%	177.5	23.81%					
7			R	208.8		-10%	187.9	23.81%					
8			$SC = 0.15R$	31.32		-5%	198.4	23.81%					
9						0%	208.8	23.81%			Note: $ATCF_k = R_k - (PC_k + SC_k + IT_k)$		
10						5%	219.2	23.81%					
11		k				10%	229.7	23.81%				Capital & $ATCF$s	CF
12		0	$LC = 0.01FC$	0.696		15%	240.1	23.81%		r_c	0.40	LC	(0.696)
13		1	$\frac{1}{3}FC$	23.200		20%	250.6	23.81%				$\frac{1}{3}FC$	(23.200)
14		2	$\frac{2}{3}FC+WC$	56.840	R	SC	PC	f	$d = fFC$	$TI = R - (PC + SC + d)$	$IT = r_c TI$	$\frac{2}{3}FC+WC$	(56.840)
15		3			208.8	31.32	147	0.35	24.36	6.12	2.45	$ATCF_3$	28.032
16		4			208.8	31.32	147	0.26	18.10	12.38	4.95	$ATCF_4$	25.526
17		5			208.8	31.32	147	0.156	10.86	19.62	7.85	$ATCF_5$	22.631
18		6			208.8	31.32	147	0.1101	7.66	22.82	9.13	$ATCF_6$	21.353
19		7			208.8	31.32	147	0.1101	7.66	22.82	9.13	$ATCF_7$	21.353
20		8			208.8	31.32	147	0.0138	0.96	29.52	11.81	$ATCF_8$	18.672
21		9			208.8	31.32	147	0	0.00	30.48	12.19	$ATCF_9$	18.288
22		10			208.8	31.32	147	0	0.00	30.48	12.19	$ATCF_{10}$	18.288
23		11	$NRC = LC + WC + (1 - r_c)S$		208.8	31.32	147	0	0.00	30.48	12.19	$ATCF_{11}$	18.288
24		12	13.224		208.8	31.32	147	0	0.00	30.48	12.19	$NRC + ATCF_{12}$	31.512
25		13		TOTALS	2,088	313.2	1,470	1.0000	69.60	235.20	94.08	= IRR($M$12:$M$24)	23.81%

Spreadsheet 10.2 ***Evaluated IRR% versus R Data Table for the Chemical Plant Investment***

	A	B	C	D	E	F	G	H	I	J	K	L	M
1													
2				$\$10^6$	*(IRR% versus R)* $_{evaluated}$ *of the CHEMICAL PLANT INVESTMENT: CT=2a, MACRS*								
3			*FC*	69.6			**Data Table**						
4			*WC* = 0.15*FC*	10.44		$\Delta R\%$	***R***	**23.81%**					
5			*S* = 0.05*FC*	3.48		-20%	**167.0**	**19.05%**					
6			*PC*	147		-15%	**177.5**	**20.24%**					
7			***R***	208.8		-10%	**187.9**	**21.43%**					
8			*SC* =0.15*R*	31.32		-5%	**198.4**	**22.62%**					
9						0%	**208.8**	**23.81%**			Note: $ATCF_k = R_k - (PC_k + SC_k + IT_k)$		
10						5%	**219.2**	**25.00%**					
11		k				10%	**229.7**	**26.19%**				Capital & *ATCF* s	*CF*
12		0	*LC* = 0.01*FC*	0.696		15%	**240.1**	**27.39%**		r_c	0.40	*LC*	(0.696)
13		1	$\frac{1}{3}FC$	23.200		20%	**250.6**	**28.58%**				$\frac{1}{3}FC$	(23.200)
14		2	$\frac{2}{3}FC+WC$	56.840	***R***	*SC*	*PC*	f	$d = fFC$	$TI = R - (PC + SC + d)$	$IT = r_c TI$	$\frac{2}{3}FC+WC$	(56.840)
15		3			208.8	31.32	147	0.35	24.36	6.12	2.45	$ATCF_3$	28.032
16		4			208.8	31.32	147	0.26	18.10	12.38	4.95	$ATCF_4$	25.526
17		5			208.8	31.32	147	0.156	10.86	19.62	7.85	$ATCF_5$	22.631
18		6			208.8	31.32	147	0.1101	7.66	22.82	9.13	$ATCF_6$	21.353
19		7			208.8	31.32	147	0.1101	7.66	22.82	9.13	$ATCF_7$	21.353
20		8			208.8	31.32	147	0.0138	0.96	29.52	11.81	$ATCF_8$	18.672
21		9			208.8	31.32	147	0	0.00	30.48	12.19	$ATCF_9$	18.288
22		10			208.8	31.32	147	0	0.00	30.48	12.19	$ATCF_{10}$	18.288
23		11	$NRC = LC + WC + (1 - r_c)S$		208.8	31.32	147	0	0.00	30.48	12.19	$ATCF_{11}$	18.288
24		12	13.224		208.8	31.32	147	0	0.00	30.48	12.19	$NRC + ATCF_{12}$	31.512
25		13		**TOTALS**	2,088	313.2	1,470	1.0000	69.60	235.20	94.08	**= IRR($M$12:$M$24)**	**23.81%**

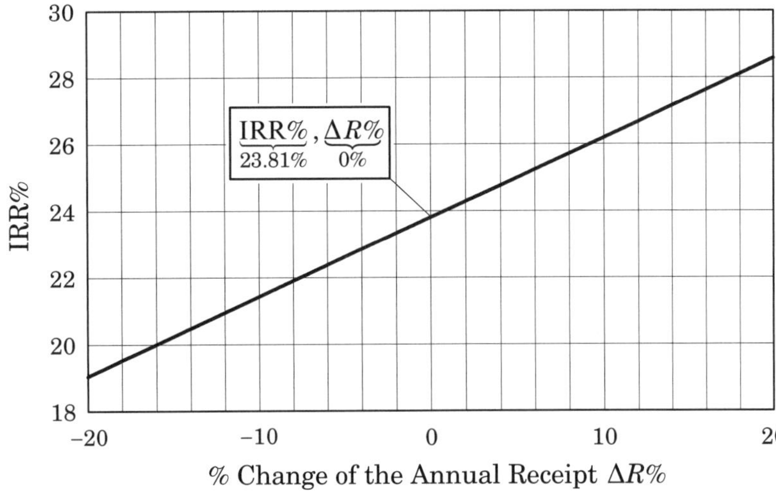

Figure 10.1 **Sensitivity Curve of the CHEMICAL PLANT INVESTMENT: *CT* = 2a, MACRS** NOTE: It shows the variation of the *IRR%* with the *percentage change* of the annual receipt $\Delta R\%$ relative to the *default value* of R: $\$208.8 \times 10^6$.

10.2 STAGED CONSTRUCTION & BREAK-EVEN POINT

Examples 4.4 and 5.3 calculate the annual and present worths, respectively, of the construction and operating costs of one large school and the staged construction and operating costs of two smaller ones. The construction of the second school is *assumed* to occur in the sixth year. But what if it occurs in the fourth year or the eighth year? How would this affect the conclusion that the single school is the least-cost alternative? We can easily answer this question with the sensitivity analysis in Spreadsheet 10.3. It calculates of the present worth of the staged construction & operating costs versus the construction year N in which they occur.

	A	B	C	D	E	F
1						
2		***Present Worth of the Costs in Staged Construction Year N***				
3		*N*	6	***Present Worths***	**Data Table**	
4			PW_P	\$1,200,000	*N*	**\$4,047,117**
5			$PW_{P_{2nd}}$	\$902,947	3	**\$4,548,648**
6			PW_A	\$1,112,732	4	**\$4,368,687**
7			PW_G	\$179,571	5	**\$4,201,812**
8			$PW_{A_{2nd}}$	\$600,331	*6*	***\$4,047,117***
9			$PW_{G_{2nd}}$	\$51,535	7	**\$3,903,757**
10			=SUM(D4:D9)	**\$4,047,117**	8	**\$3,770,950**
11					9	**\$3,647,968**

Spreadsheet 10.3 **Data Table for the Present Worth of the Staged School Construction and Operating Costs *Versus* the Construction Year *N***

Computation Notes: *(1) Spreadsheet 10.3 is located in the Book Templates file provided with the book. Since this file contains the* ***12-FACTOR Calculator****, the 12 factor equations are available in* ***every*** *spreadsheet the Book Templates file.*
(2) The present worth values in array D4:D9 of Spreadsheet 10.3 are calculated with ***factor equations****.*
(3) The numbers in array D4:D9 are essentially those in Table 5.6, which are calculated with factors values rounded to five digits.
(4) To evaluate the Data Table we follow Step (4a) of Step-by-Step 10.1. In the **Data Table** *dialog box Click cell C3 in the* **Column input cell:** *box.*

The evaluated **Data Table** shows the variation of the present worth of the staged construction and operating costs with the construction year N. The present worth of the construction and operating costs of one school were calculated as $3.828 \times \$10^6$ in Example 5.5. Hence, ***if the staged construction occurred in the eighth year, it would be the least-cost alternative!***

Figure 10.2 compares the present worths of the costs of constructing and operating one school and two schools. The **break-even point**, the time at which the cost present worths of the alteratives are equal, is about 7.5 years. *After that time the two school alternative is the best alternative.*

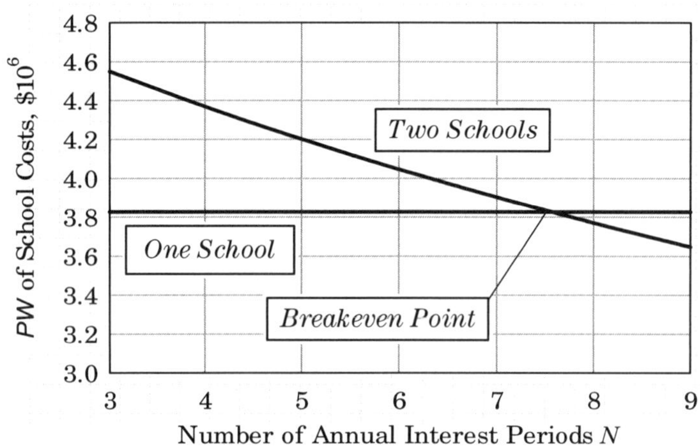

Figure 10.2 **Present worth cost comparison of constructing one school and a staged construction that satisfies the same need**

Practice Note: *The critical element in choosing between the two alternatives is the* ***accuracy*** *of the consultant's* ***estimate*** *that a second school will be required in six years (see Example 4.4). Predicting an event six years in the future is not easy. If a second school turns out to be required in less than 7.5 years, building two schools would have been be the* ***wrong*** *decision. But if the future demand for classroom space occurs* ***after*** *7.5 years, building the second school at that time would be the* ***right*** *decision.* ***Making cost comparisons is easy, predicting the future is difficult!***

10.3 SUMMARY

Step-by-Step 10.1 in Section 10.1 explains how to create a **Data Table** in Excel to determine how changing values of one variable affects the values of the result of a calculation. Spreadsheets 10.1 and 10.2 show how to create a **Data Table** to determine how changing the value of the annual receipt of the chemical plant investment affects its IRR%. Figure 10.1 displays the resulting **Sensitivity Curve**.

Section 10.2 explains how create a Data Table and Sensitivity Curve for a least-cost comparison of the fixed capital and operating costs of one large school and those of two small schools. This calculation also provides the **break-even point**, the time at which the cost present worths of the alteratives are equal. Figure 10.2 displays the break-even point for the one school versus two smaller schools alternatives.

10.4 PROBLEMS

(P10-1) *Create a **Data Table** for the IRR% versus the fixed capital FC of the **CHEMICAL PLANT INVESTMENT** in Spreadsheet 10.1,* a copy of which is in the *Book Templates* file. Let the values of *FC* vary from –20% to 20% of the base case value in 5% increments.

(P10-2) *Plot the sensitivity curves of the IRR% versus the fixed capital FC **and** the annual receipt R of the **CHEMICAL PLANT INVESTMENT**.* NOTE: ***These types of curves are essential for the investment decision maker!***

(P10-3) *Prepare a present worth break-even analysis of the potable water supply systems described in problem (P4-4).*

11

Other Profitability Measures

OBJECTIVES

Calculating the IRR of an *proposed* investment is the most popular profitability measure used in industry. However, a survey of 128 companies reveals that *more than one profitability measure is used to evaluate proposed investments*[1]*:*

1. IRR: 80%
2. Net Present Value: 78%
3. Payback Period: 52%
4. Accounting Rate of Return: 34%

After reading this chapter you will be able to calculate the profitability of an investment by its (1) *Net Present Value*, (2) *Accounting Rate of Return*, and (3) *Benefit-Cost Ratio*. In addition, you will understand the role of the *Payback Period* in evaluating a proposed investment.

11.1 NET PRESENT VALUE

11.1.1 A 15-Second Review of Calculating an IRR

We start with **General Repayment Equation**:

$$\boxed{\sum_{k=0}^{N} \frac{R_k - D_k}{(1+i^*)^k} = \sum_{k=0}^{N} \frac{CF_k}{(1+i^*)^k} = 0.} \tag{2-107}$$

It leads to the **General Polynomial Equation** (GPE):

$$\boxed{CF_0 x^N + CF_1 x^{N-1} + \cdots + CF_{N-1} x + CF_N = 0,} \tag{2-111}$$

where

$$\boxed{i^* \equiv x_+ - 1,} \tag{2-110}$$

and x_+ *is the unique positive root* of the GPE *of a simple investment.* Finding the *value of* i^* is easy with a personal computer and Excel.

[1]Farragher, E.J., Kleiman, R.T. & Sahu, A.P. *The Engineering Economist*, 44:2, 137-150, 1999. The percentage following a profitability measure is the percentage of the companies using the method.

11.1.2 Screening the IRR with a Net Present Value Calculation

Calculating the *pretax* or *aftertax* IRR of a *simple* investment *without* a personal computer and mathematics software is *tedious*. Hence, before they were available in the late eighties, a simple investment was usually ***screened*** for its IRR. This involves first calculating its **net present value** (NPV) at the **minimum acceptable rate of return (MARR)** r^*:

$$NPV(r^*) \equiv PW(r^*) = \sum_{k=0}^{N} \frac{CF_k}{(1+r^*)^k}. \qquad (11\text{-}1)$$

The following relationships determine if i^* is greater than, equal to, or less than r^*:

$$\begin{array}{|ll|} \hline NPV(r^*) > 0 & i^* > r^* \\ NPV(r^*) = 0 & i^* = r^* \\ NPV(r^*) < 0 & i^* < r^*. \\ \hline \end{array} \qquad (11\text{-}2)$$

Figure 11.1 illustrates these $NPV(r^*)$ relationships.

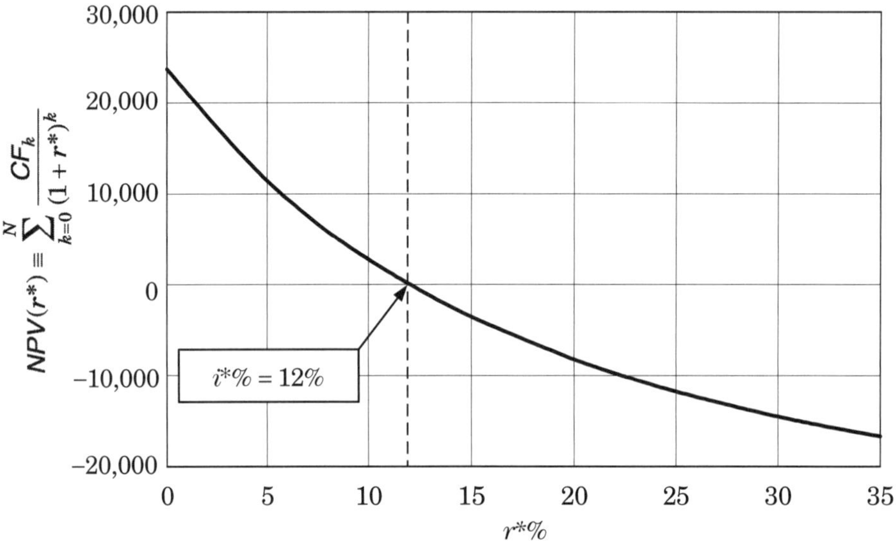

Figure 11.1 **Variation of the NPV with $r^*\%$ for a *Simple* Investment**

11.1.3 Calculating a Net Present Value with Excel

We use Excel's NPV function to calculate an NPV. Its syntax is NPV(r^*, *array*) where r^* is the minimum acceptable rate of return, and *array* is a column or row array of the investment's cash flows ***except its first cash flow. It must be added to the result.*** See label cell D17 of Example 11.1.

EXAMPLE 11.1 Calculating a Pretax NPV **of the** Chemical Plant Investment

Problem: *Calculate the* ***pretax*** *NPV of the chemical plant investment in Spreadsheet 6.1 when* $r^*\% = 30\%$.

Solution: Spreadsheet 11.1 provides the solution. Since the value of NPV is positive, $i^*\% > 30\%$. Spreadsheet 6.1 shows that $i^*\% = 32.59\%$.

	A	B	C	D	E
1					
2		$r^*\%$	30%	***Pretax NPV of the Chemical Plant Investment***	
3			k		CF, \$$10^6$
4			0	$-LC$	(0.696)
5			1	$-\frac{1}{3}FC$	(23.2)
6			2	$-\left(\frac{2}{3}FC+WC\right)$	(56.84)
7			3		30.48
8			4		30.48
9			5		30.48
10			6		30.48
11			7	OCF	30.48
12			8		30.48
13			9		30.48
14			10		30.48
15			11		30.48
16			12	$OCF+RC$	45.096
17				=NPV(C2,E5:E16)+E4	**4.2095**

Spreadsheet 11.1 ***Pretax*** *NPV* of the Chemical Plant Investment

11.1.4 Comparing Two or More Mutually Exclusive Investments

To compare n mutually exclusive investments (MEIs) by their net present values, we use Step-by-Step 11.1.

General MEI Selection Procedure by *NPV*: Step-by-Step 11.1

(1) Order the cash flows of the n MEIs by their *increasing* total capitals and calculate their *NPV*s with Excel's NPV function *plus* the first cash flow. If *all* of the *NPV*s are *negative*, *reject them*; otherwise go to Step 2.

(2) Make the *first* MEI whose *NPV* is nonnegative the **defender** $\mathbb{D}$ and the *next* MEI whose *NPV* is nonnegative the challenger $\mathbb{C}$; go to Step 3. If there is no challenger, select the **defender**.

(3) Calculate the *NPV* of the incremental investment of $\mathbb{C}$ versus $\mathbb{D}$: $NPV_{\mathbb{C}-\mathbb{D}}$.

(4A) If $NPV_{\mathbb{C}-\mathbb{D}} \geq 0$, make the challenger the defender and determine if a new challenger exists. If it does, repeat Step 3. If a new challenger doesn't exist, select the defender.

(4B) If $NPV_{\mathbb{C}-\mathbb{D}} < 0$, determine if a new challenger exists. If it does, go to Step 3. If a new challenger doesn't exist, select the defender.

Spreadsheet 11.2 applies Step-by-Step 11.1 to a *NPV* comparison of the six MEIs that were previously compared by their ***aftertax*** IRRs in Example 8.2.

	A	B	C	D	E	F	G	H
1								
2		***Aftertax* NPV *Comparison of 6 MEIs: r*% = 20%***						
3		***k***	CF_U	CF_V	CF_W	CF_X	CF_Y	CF_Z
4		0	(1)	(1.2)	(1.3)	(1.5)	(1.6)	(2)
5		1	(20)	(22)	(25)	(26)	(27)	(30)
6		2	(40)	(42)	(43)	(44)	(47)	(51)
7		3	17	18	19	18	22	24
8		4	19	20	21	20	24	26
9		5	21	22	23	22	26	28
10		6	23	24	25	24	28	30
11		7	25	26	27	26	30	32
12		8	27	28	29	28	32	34
13		9	27	28	29	28	32	34
14		10	27	28	29	28	32	34
15		11	27	28	29	28	32	34
16		12	34	35.6	37.1	36.5	41	44.1
17		**NPV**	**20.00**	**19.72**	**19.39**	**14.80**	**23.48**	**23.75**
18		***Calculation of the NPVs of the Incremental Investments***						
19		*k*	CF_{C-D}	CF_{V-U}	CF_{W-U}	CF_{X-U}	CF_{Y-U}	CF_{Z-Y}
20		0		(0.2)	(0.3)	(0.5)	(0.6)	(0.4)
21		1		(2)	(5)	(6)	(7)	(3)
22		2		(2)	(3)	(4)	(7)	(4)
23		3		1	2	1	5	2
24		4		1	2	1	5	2
25		5		1	2	1	5	2
26		6		1	2	1	5	2
27		7		1	2	1	5	2
28		8		1	2	1	5	2
29		9		1	2	1	5	2
30		10		1	2	1	5	2
31		11		1	2	1	5	2
32		12		1.6	3.1	2.5	7	3.1
33			**NPV**	**(0.28)**	**(0.60)**	**(5.20)**	**3.49**	**0.27**
34			*C*	V	W	X	Y	**Z**
35			*D*	*U*	*U*	*U*	*U*	Y

Spreadsheet 11.2 ***Aftertax Net Present Value Comparison of the Six Pharmaceutical Plant MEIs*** As in Example 8.2, *MEI Z is the best investment.* NOTES: (1) Each value of *NPV* is calculated as NPV (20%, CF_{2nd}:CF_{last}) + CF_{1st}. (2) The top and bottom halves of the spreadsheet calculate the *NPV*s of the *aftertax* cash flows of the investments and the incremental investments, respectively.

Spreadsheets 11.1 and 11.2 are in the Book Templates file.

11.2 ACCOUNTING RATE of RETURN (*ARR*)

The **accounting rate of return** (*ARR*) is *defined* as the *average* accounting profit (*AP*) divided by the investment's fixed capital[2]:

$$ARR \equiv \frac{AP}{FC}, \tag{11-3}$$

where the ***average* accounting profit** (*AP*) is *defined* as the difference of the *estimated* ***pretax*** operating cash flows and the depreciations divided by the operating time of the investment:

$$AP \equiv \frac{1}{OT}\sum_{k^*=1}^{OT}(OCF_{k^*} - d_{k^*}), \tag{11-4}$$

where

- *OCF_{k^*} is the operating cash flow in the k*th year of the operating time.*
- d_{k^*} is the depreciation charge for the k^*th *year of the operating time.*
- OT is the *assumed* operating time of the investment measured in annual interest periods.

The value of d_{k^*} is calculated with **straight line depreciation**[3]:

$$d_{k^*} \equiv \frac{FC - S}{OT}. \tag{11-5}$$

EXAMPLE 11.2 Calculating the *ARR* of the Chemical Plant Investment

Problem: *Calculate the ARR of the chemical plant investment whose pretax IRR was previously calculated in Spreadsheet 6.1.*

Solution: Spreadsheet 6.1 shows that

- $FC = \$69.60 \times 10^6$
- $S = \$3.48 \times 10^6$
- $OT = 10$ annual interest periods
- $OCF_k = \$30.48 \times 10^6$

We first calculate the value of d_{k^*} with Equation (11-5):

$$d_{k^*} = \frac{FC - S}{OT} = \frac{(69.60 - 3.48)\times 10^6}{10} = \$6.61 \times 10^6.$$

Next the average accounting profit (*AP*) is calculated with Equation (11-4):

---

[2] Implicit in the definition of the *ARR* is the *assumption* that the working capital and land capital (if any) are fully recovered.

[3] NOTES: (1) We do not use MACRS depreciation. (2) A salvage or market value can be used in Equation (11-5).

$$AP = \frac{1}{OT}\sum_{k^*=1}^{OT}(OCF_{k^*} - d_{k^*}) = \frac{1}{10}\sum_{k^*=1}^{10}(\$30.48\times10^6 - \$6.61\times10^6) = \$23.87\times10^6.$$

Hence, it follows from Equation (11-3) that

$$ARR\% \equiv 100\frac{AP}{FC} = 100\frac{\$23.87\times10^6}{\$69.60\times10^6} = \underline{\underline{34.30\%}}.$$

This result is close to the *pretax* IRR% of 32.59% calculated in Spreadsheet 6.1.

Computation Note: *The result of Example 11.2 and its "agreement" with a corresponding pretax IRR% calculation shows why some companies use accounting rate of return (ARR) and the IRR% in pretax profitability calculations.*

11.3 BENEFIT-COST RATIO

For any *simple* investment, we can rewrite the General Repayment Equation as

$$\sum_{k=0}^{N}\frac{R_k}{(1+i^*)^k} = \sum_{k=0}^{N}\frac{D_k}{(1+i^*)^k}, \tag{11-6}$$

where $\boldsymbol{i^*}$ is the **rate of return**. If we divide both sides of the equation by its right-hand side we obtain the following expression:

$$\frac{B}{C}(i^*) \equiv \frac{\sum_{k=0}^{N}\frac{R_k}{(1+i^*)^k}}{\sum_{k=0}^{N}\frac{D_k}{(1+i^*)^k}} = 1, \tag{11-7}$$

where the function $\frac{B}{C}(i^*)$ is called the **benefit-cost (*B-C*) ratio**. It is defined as *the ratio of the present worth of the receipts (benefits) and the disbursements (costs) of a simple investment.*

We use the following relationships to determine if i^* is greater than, equal to, or less than r^*:

$$\begin{aligned} &\frac{B}{C}(r^*) > 1 \qquad i^* > r^* \\ &\frac{B}{C}(r^*) = 1 \qquad i^* = r^* \\ &\frac{B}{C}(r^*) < 1 \qquad i^* < r^*. \end{aligned} \tag{11-8}$$

Practice Notes: *(1) To be considered for funding, a proposed investment must have a B-C ratio ≥ 1 at the specified MARR (r^*).*

(2) Analyzing an investment by its B-C ratio is not common in industry. However, it is normally used in the U.S. government to analyze the multitude of projects that it considers for funding. I explain it so that you are familiar with it. Who knows, you may work for the U.S. government or be a consultant to it.

EXAMPLE 11.3 Calculating a Pretax *B-C* Ratio of the Chemical Plant Investment

Problem: *Calculate the pretax B-C Ratio of the chemical plant investment when $r^*\% = 30\%$.*

Solution: Spreadsheet 11.3 shows the easy calculation in Excel.

	A	B	C	D	E	F	G	H
1								
2		$r^*\%$	30%	***Pretax B-C Ratio of the Chemical Plant***				
3			***k***		CF_B, $\$10^6$	CF_C, $\$10^6$		
4			0		0	0.696	$-LC$	
5			1		0	23.2	$-\frac{1}{3}FC$	
6			2		0	56.84	$-\left(\frac{2}{3}FC+WC\right)$	
7			3		30.48			
8			4		30.48			
9			5		30.48			
10			6		30.48			
11			7	*OCF*	30.48		$\frac{B}{C}(30\%) =$	**1.081**
12			8		30.48			
13			9		30.48			
14			10		30.48			
15			11		30.48			
16			12	*OCF + RC*	45.096			
17			=NPV(C2,E5:E16)+E4		**56.3848**	**52.1753**	=NPV(C2,F5:F6)+F4	
18					***B***	***C***		

Spreadsheet 11.3 ***Pretax Benefit-Cost Ratio of the Chemical Plant Investment*** NOTE: *This spreadsheet is in the Book Templates file.* Since the B-C ratio is greater than 1, the IRR% is greater than 30%. If we use Goal Seek to find the value of $r^*\%$ that reduces the *B-C* ratio to 1, it returns 32.59%, which is the value of $i^*\%$ calculated in Spreadsheet 6.1.

11.4 PAYBACK PERIOD

11.4.1 Definition & Calculation

The **payback period** *PP* of an investment is the number of interest periods required to recover the sum of its *capital* cash flows from its positive cash flows. For example, consider the following *expected* cash flows of a cost reduction investment:

CASH FLOW, $\$10^3$	−15	−2	5	5	5	4	4	4	4	4	4
YEAR k	0	1	2	3	4	5	6	7	8	9	10

The following **Payback Table** presents the calculation of PP for this investment: *

Payback Table for the Cost Reduction Investment

YEAR	POSITIVE CF, $\$10^3$	CUMULATIVE CF, $\$10^3$
2	5	5
3	5	10
4	5	15
5	4	**17 + 2**

The ***cumulative*** payback cash flow of $17,000 is received in the *middle* of the 5th year. Hence,

$$PP = 3.5\text{ a.}$$

Computation Note:*(1) A payback period calculation does not consider the time value of the investment's cash flows.*
(2) A payback period calculation does not consider any cash flows after the last payback cash flow. Hence, ***it does not consider the profitability of the investment!***
(3) ***The goal of a payback period calculation is to determine how long it takes to recover an investment's capital cash flows!***

11.4.2 Rationale for the Payback Period

A payback period calculation

- ignores the time value of money.
- ignores all cash flows after the payback period.
- ignores the expected profitability of the investment.

Hence, it's reasonable to ask a simple question. ***"What does a payback period calculation provide?"***

The **payback period** is a measure of

1. *the risk of not recovering some or all of the investment's capital.*
2. *the effect of an investment's capital on the capital available for funding other projects.*

For example, there are government and competition ***risks*** associated with many investments. Hence, a large payback period means that some or all of the investment's capital may be lost if it is terminated because of *new* environmental regulations, or the investment can't compete with the product pricing of a *new* competitor.

In addition, companies do not have unlimited sources of capital available to fund new investments. Hence, a small payback period makes an investment's capital available to fund or partially fund a new investment.

It should be clear from the preceding that ***the payback period is a capital***

budgeting tool and not an investment tool! The decision to fund a project will normally depend on its profitability—as measured by its IRR% or Net Present Value—*and* the *effect* of the project's payback period on the firm's available investment capital. A notable exception to this conclusion are companies that manufacture consumer electronics. They may *primarily* use the payback period to evaluate new investments because the rapid changes in their product technologies often lead to short product lives of several years.

EXAMPLE 11.4 Calculating the Aftertax Payback Period of the Chemical Plant Investment

Problem: Calculate the *aftertax* **payback period** of the chemical plant investment in Spreadsheet 9.1.

Solution: Spreadsheet 11.4 displays the capital and aftertax cash flows and the calculation of the IRR%.

	A	B	C	D
1				
2		***Chemical Plant Aftertax Cash Flows and IRR%***		
3		***k***	***Capital & ATCFs***	***CF*, $10⁶**
4		0	LC	(0.696)
5		1	$\frac{1}{3}FC$	(23.2)
6		2	$\frac{2}{3}FC+WC$	(56.84)
7		3	$ATCF_3$	28.032
8		4	$ATCF_4$	25.526
9		5	$ATCF_5$	22.631
10		6	$ATCF_6$	21.353
11		7	$ATCF_7$	21.353
12		8	$ATCF_8$	18.672
13		9	$ATCF_9$	18.288
14		10	$ATCF_{10}$	18.288
15		11	$ATCF_{11}$	18.288
16		12	$NRC + ATCF_{12}$	31.512
17		13	**=IRR(D4:D16)**	**23.81%**

Spreadsheet 11.4 ***Aftertax IRR% of the Chemical Plant Investment***

The **total capital** *TC* is given by

$$TC = (0.696 + 23.2 + 56.84) \times \$10^6 = 80.736 \times \$10^6.$$

The Payback Table on the following page presents the calculation of *PP*. At the *end* of the fifth year we will need $(80.736 - 76.189) \times 10^6 = \$4.547 \times 10^6$ from the $\$21.353 \times 10^6$ operating cash flow for the sixth year to recover the total capital.

Aftertax Payback Table for the Chemical Plant Investment

YEAR	OPERATING CF, $\$10^6$	CUMULATIVE CF, $\$10^6$
3	28.032	28.032
4	25.526	53.558
5	22.631	76.189
6	21.353	**80.736** + 16.806

Hence, the payback period for the investment is 3.21 a:

$$PP = 3 + \frac{4.547}{21.353} = \underline{\underline{3.21\text{ a.}}}$$

11.5 INVESTMENT INFORMATION PACKAGE

An **Investment Information Package** (IIP) for any proposed manufacturing plant contains the following items:

1. Narrative describing the investment.
2. Aftertax IRR spreadsheet with MACRS depreciation.
3. Aftertax sensitivity tables for the fixed capital, annual production costs, and annual receipts, and a graph that displays the data.
4. Payback period calculation.
5. Environmental impact assessment.

The IIP provides *some* of the information that a company's management requires to make a decision to fund a proposed investment. ***Marketing information***, such as competitors, product pricing, potential market share, and competing products will also affect the decision.

11.6 SUMMARY

Section 11.1 explains the use of a **net present value** NPV calculation for screening the *pretax* or *aftertax* profitability of an investment. Example 11.1 calculates a *pretax* NPV of the chemical plant investment. Step-by-Step 11.1 explains how to compare two or more mutually exclusive investments (MEIs) by their NPVs, and Spreadsheet 11.2 illustrates an *aftertax* NPV comparison of six MEIs.

Section 11.2 explains the **accounting rate of return** (ARR) for estimating the *pretax* profitability of an investment, and Example 11.2 illustrates its use by calculating the ARR of the chemical plant investment.

Section 11.3 explains the **benefit-cost ratio** (B-C) ratio and shows how to use its value for a specified value of r^* to determine whether an investment's i^* is greater than, equal to, or less than r^*. Example 11.3 illustrates its use by calculating a *pretax* B-C ratio of the chemical plant investment.

Section 11.4 defines the **payback period** of an investment and explains its role in the investment selection process. Example 11.4 illustrates its use by calculating the *aftertax* **payback period** of the chemical plant investment.

Section 11.5 describes the contents of an **Investment Information Package**.

11.7 PROBLEMS

(P11-1) (Part A) *Calculate the aftertax NPV of the investment in Example 3.5 when r*% is 20%.* **(Part B)** *Explain in one very short sentence what the result means.* **(Part C)** *Use Goal Seek to find the value of r*% that makes NPV = 0.* **(Part D)** *What do we call this value of r*%?*

(P11-2) *Calculate the accounting rate of return of an investment with the following aftertax cash flows:*

CF, $10^4	−4	−3	1	2	3	3	3	2
YEAR *k*	0	1	2	3	4	5	6	7

NOTE: Twenty-five percent of the first negative cash flow represents working capital.

(P11-3) (Part A) *Calculate the aftertax* B-C ratio *of the investment in Example 3.5 when r*% is 30%.* **(Part B)** *Explain in one very short sentence what the result means.* **(Part C)** *Use Goal Seek to find the value of r*% that makes* $\frac{B}{C} = 1$. **(Part D)** *What do we call this value of r*%?*

(P11-4) *Calculate the aftertax payback period of the investment in Example 3.5.*

(P11-5) (Part A) *Use the cash flows in Spreadsheet 11.4 to find the aftertax NPV when r*% is 25%.* **(Part B)** *Use Goal Seek to find the value of r*% that makes NPV = 0.* **(Part C)** *What do we call this value of r*%?*

INDEX

N

O

P

Q

R

S

T

U

W